Number Grids

Games and Activities to Investigate Numbers

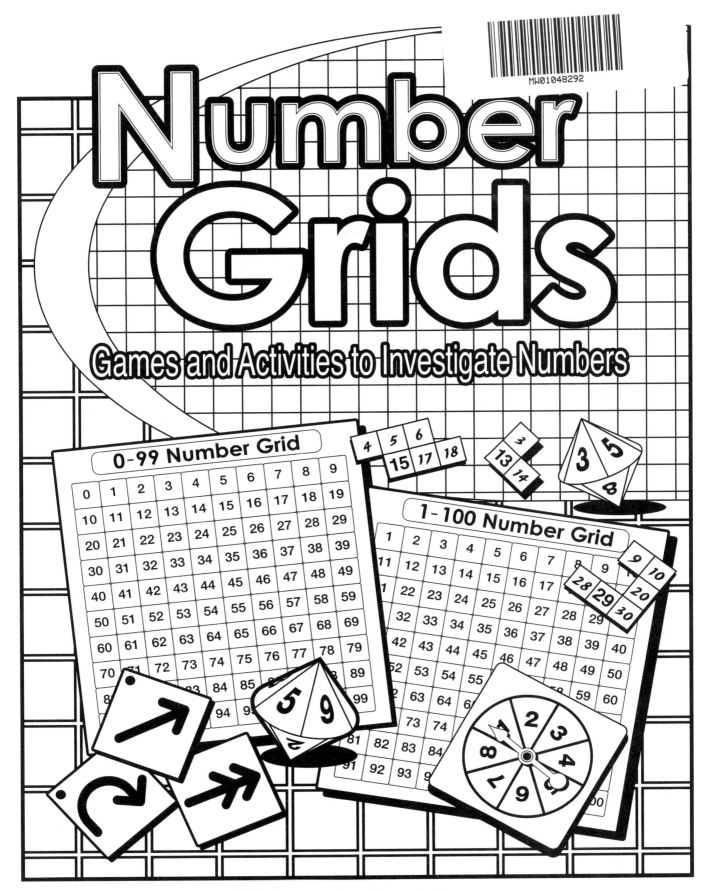

Written by Paul Swan

Published by Didax Educational Resources

www.didaxinc.com

Published with the permission of R.I.C. Publications Pty. Ltd.

First published by R.I.C. Publications Pty. Ltd., Perth, Western Australia. Revised by Didax Educational Resources.

Printed in the United States of America.

Order Number 2-163
ISBN 1-58324-160-4

A B C D E F 07 06 05 04 03

Educational Resources
395 Main Street
Rowley, MA 01969
www.didaxinc.com

Foreword

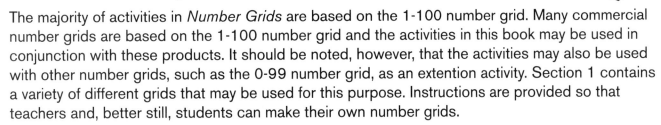

Number grids are an inexpensive and versatile teaching aid that may be used:

- to develop number skills,
- as a means of examining number patterns, and
- as a source of investigation.

The majority of activities in *Number Grids* are based on the 1-100 number grid. Many commercial number grids are based on the 1-100 number grid and the activities in this book may be used in conjunction with these products. It should be noted, however, that the activities may also be used with other number grids, such as the 0-99 number grid, as an extention activity. Section 1 contains a variety of different grids that may be used for this purpose. Instructions are provided so that teachers and, better still, students can make their own number grids.

Number grids may be used on a regular basis to assist in the development of mental computation skills. The possible uses of number grids are unlimited when students are given the chance to explore.

Contents

How to Use This Book

Number Grids is divided into three sections.

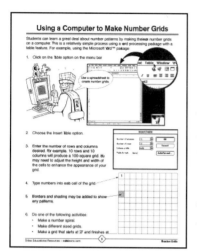

Section 1 contains information about how students can make their own number grids using a computer.

It also contains a variety of completed number grids and blank grids to photocopy for students to use with the games and activities in Section 3. These can be enlarged, photocopied on card stock and laminated. Students may use a water-soluble marker to practice an activity before completing the associated worksheet in Section 3.

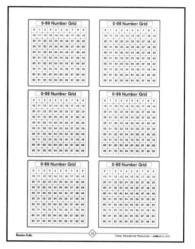

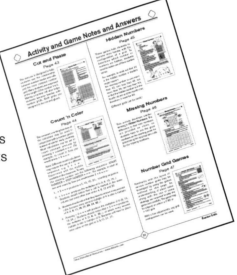

Section 2 contains detailed background information about each game and activity in Section 3. The purpose of each game and activity is explained, along with teaching points and ideas to assist students' understanding. Answers are also provided where necessary.

Section 3 contains a collection of games and activities. Some need to be photocopied for each student. Others can be photocopied once and then laminated or placed in a plastic sleeve for student use.

Other equipment needed by the students such as dice, cards and cubes is listed on the appropriate pages.

The games and activities are arranged in order of difficulty.

The majority of games and activities use the 1-100 number grid as an example. However, in most cases, any grid configuration can be substituted.

Background Information

A **number grid** is a convenient way of representing a set of numbers. A grid is made up of **columns** and **rows**.

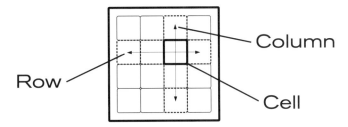

The intersection of a column and a row is called a **cell**.

Traditionally, the numbers have been laid out in a 10x10 grid of 100 cells. Typically, the numbers are laid out from 0-99 or 1-100.

1	2	3	4	5	6	7	8	9	10
11	12	13	14	15	16	17	18	19	20
21	22	23	24	25	26	27	28	29	30
31	32	33	34	35	36	37	38	39	40
41	42	43	44	45	46	47	48	49	50
51	52	53	54	55	56	57	58	59	60
61	62	63	64	65	66	67	68	69	70
71	72	73	74	75	76	77	78	79	80
81	82	83	84	85	86	87	88	89	90
91	92	93	94	95	96	97	98	99	100

0	1	2	3	4	5	6	7	8	9
10	11	12	13	14	15	16	17	18	19
20	21	22	23	24	25	26	27	28	29
30	31	32	33	34	35	36	37	38	39
40	41	42	43	44	45	46	47	48	49
50	51	52	53	54	55	56	57	58	59
60	61	62	63	64	65	66	67	68	69
70	71	72	73	74	75	76	77	78	79
80	81	82	83	84	85	86	87	88	89
90	91	92	93	94	95	96	97	98	99

There is some debate as to how the configuration of the grid and the numbers within the grid might affect a student's understanding of number. For example, it has been suggested that a 0-99 grid has the advantage of heightening a student's awareness of zero. The 0-99 grid also has the advantage of a new decade beginning each row.

Those who favor the 1-100 configuration believe students become confused when counting on a 0-99 grid. Often students will count the first square as "one" even though it is labeled zero. This means the count and the number of squares will always differ by one.

Another school of thought suggests that a 1-99 grid be used, where the zero is removed from the corner of the grid. This format has the advantage of each decade beginning a new row and each square matching the count.

Rather than focus on a single representation or grid, a variety of grids have been provided in Section 1 of this book. For example, instead of starting a grid at one or zero, why not start at 27? Why not try different-sized grids, not just the traditional 10x10, one-hundred-cell grid? Not all number grids need to be 10 rows high and 10 columns wide. Neither do grids have to start at zero or one; they can start at any number and do not have to step up in ones. Another suggestion is to have students create their own grids. Instructions are provided in Section 1 to help students make use of technology to produce a variety of number grids.

	1	2	3	4	5	6	7	8	9
10	11	12	13	14	15	16	17	18	19
20	21	22	23	24	25	26	27	28	29
30	31	32	33	34	35	36	37	38	39
40	41	42	43	44	45	46	47	48	49
50	51	52	53	54	55	56	57	58	59
60	61	62	63	64	65	66	67	68	69
70	71	72	73	74	75	76	77	78	79
80	81	82	83	84	85	86	87	88	89
90	91	92	93	94	95	96	97	98	99

Number Grid Templates
pages 7-27

In this section you will find:

1. Information about how students can make their own number grids using a computer.

2. A variety of completed number grids and blank grids to photocopy for students to use with the games and activities in Section 3. The grids can be enlarged and laminated for durability. Laminating the grids allows students to write on them with a water-soluble marker to test out their ideas. The marking can then be erased so the grids may be used again later.

3. A set of arrow cards to use with **Grid Race** on page 56 in Section 3.

S
e
c
t
i
o
n
1

Number Grids

Using a Computer to Make Number Grids

Students can learn a great deal about number patterns by making their own number grids on a computer. This is a relatively simple process using a word processing package with a table feature. For example, using the Microsoft Word™ package:

1. Click on the Table option on the menu bar.

Use a spreadsheet to create number grids.

2. Choose the Insert Table option.

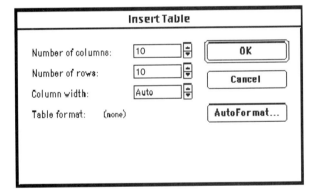

3. Enter the number of rows and columns desired. For example, 10 rows and 10 columns will produce a 100-square grid. You may need to adjust the height and width of the cells to enhance the appearance of your grid.

4. Type numbers into each cell of the grid.

5. Borders and shading may be added to show any patterns.

6. Do one of the following activities:
    *   Make a number spiral.
    *   Make different sized grids.
    *   Make a grid that starts at 27 and finishes at ____.

0-99 Number Grid

0	1	2	3	4	5	6	7	8	9
10	11	12	13	14	15	16	17	18	19
20	21	22	23	24	25	26	27	28	29
30	31	32	33	34	35	36	37	38	39
40	41	42	43	44	45	46	47	48	49
50	51	52	53	54	55	56	57	58	59
60	61	62	63	64	65	66	67	68	69
70	71	72	73	74	75	76	77	78	79
80	81	82	83	84	85	86	87	88	89
90	91	92	93	94	95	96	97	98	99

1-100 Number Grid

1	2	3	4	5	6	7	8	9	10
11	12	13	14	15	16	17	18	19	20
21	22	23	24	25	26	27	28	29	30
31	32	33	34	35	36	37	38	39	40
41	42	43	44	45	46	47	48	49	50
51	52	53	54	55	56	57	58	59	60
61	62	63	64	65	66	67	68	69	70
71	72	73	74	75	76	77	78	79	80
81	82	83	84	85	86	87	88	89	90
91	92	93	94	95	96	97	98	99	100

1-99 Number Grid

	1	2	3	4	5	6	7	8	9
10	11	12	13	14	15	16	17	18	19
20	21	22	23	24	25	26	27	28	29
30	31	32	33	34	35	36	37	38	39
40	41	42	43	44	45	46	47	48	49
50	51	52	53	54	55	56	57	58	59
60	61	62	63	64	65	66	67	68	69
70	71	72	73	74	75	76	77	78	79
80	81	82	83	84	85	86	87	88	89
90	91	92	93	94	95	96	97	98	99

Multiple Grid

1	2	3	4	5	6	7	8	9	10
11	12	13	14	15	16	17	18	19	20
21	22	23	24	25	26	27	28	29	30
31	32	33	34	35	36	37	38	39	40
41	42	43	44	45	46	47	48	49	50
51	52	53	54	55	56	57	58	59	60
61	62	63	64	65	66	67	68	69	70
71	72	73	74	75	76	77	78	79	80
81	82	83	84	85	86	87	88	89	90
91	92	93	94	95	96	97	98	99	100

100 Square Grid

2-200 Number Grid

2	4	6	8	10	12	14	16	18	20
22	24	26	28	30	32	34	36	38	40
42	44	46	48	50	52	54	56	58	60
62	64	66	68	70	72	74	76	78	80
82	84	86	88	90	92	94	96	98	100
102	104	106	108	110	112	114	116	118	120
122	124	126	128	130	132	134	136	138	140
142	144	146	148	150	152	154	156	158	160
162	164	166	168	170	172	174	176	178	180
182	184	186	188	190	192	194	196	198	200

27-126 Number Grid

27	28	29	30	31	32	33	34	35	36
37	38	39	40	41	42	43	44	45	46
47	48	49	50	51	52	53	54	55	56
57	58	59	60	61	62	63	64	65	66
67	68	69	70	71	72	73	74	75	76
77	78	79	80	81	82	83	84	85	86
87	88	89	90	91	92	93	94	95	96
97	98	99	100	101	102	103	104	105	106
107	108	109	110	111	112	113	114	115	116
117	118	119	120	121	122	123	124	125	126

1-100 Zig Zag Grid

1	2	3	4	5	6	7	8	9	10
20	19	18	17	16	15	14	13	12	11
21	22	23	24	25	26	27	28	29	30
40	39	38	37	36	35	34	33	32	31
41	42	43	44	45	46	47	48	49	50
60	59	58	57	56	55	54	53	52	51
61	62	63	64	65	66	67	68	69	70
80	79	78	77	76	75	74	73	72	71
81	82	83	84	85	86	87	88	89	90
100	99	98	97	96	95	94	93	92	91

Snakes & Ladders Grid

100	99	98	97	96	95	94	93	92	91
81	82	83	84	85	86	87	88	89	90
80	79	78	77	76	75	74	73	72	71
61	62	63	64	65	66	67	68	69	70
60	59	58	57	56	55	54	53	52	51
41	42	43	44	45	46	47	48	49	50
40	39	38	37	36	35	34	33	32	31
21	22	23	24	25	26	27	28	29	30
20	19	18	17	16	15	14	13	12	11
1	2	3	4	5	6	7	8	9	10

Diagonal 1-100 Grid

1	2	4	7	11	16	22	29	37	46
3	5	8	12	17	23	30	38	47	56
6	9	13	18	24	31	39	48	57	65
10	14	19	25	32	40	49	58	66	73
15	20	26	33	41	50	59	67	74	80
21	27	34	42	51	60	68	75	81	86
28	35	43	52	61	69	76	82	87	91
36	44	53	62	70	77	83	88	92	95
45	54	63	71	78	84	89	93	96	98
55	64	72	79	85	90	94	97	99	100

Down and Up Grid

1	20	21	40	41	60	61	80	81	100
2	19	22	39	42	59	62	79	82	99
3	18	23	38	43	58	63	78	83	98
4	17	24	37	44	57	64	77	84	97
5	16	25	36	45	56	65	76	85	96
6	15	26	35	46	55	66	75	86	95
7	14	27	34	47	54	67	74	87	94
8	13	28	33	48	53	68	73	88	93
9	12	29	32	49	52	69	72	89	92
10	11	30	31	50	51	70	71	90	91

Up and Down Grid

10	11	30	31	50	51	70	71	90	91
9	12	29	32	49	52	69	72	89	92
8	13	28	33	48	53	68	73	88	93
7	14	27	34	47	54	67	74	87	94
6	15	26	35	46	55	66	75	86	95
5	16	25	36	45	56	65	76	85	96
4	17	24	37	44	57	64	77	84	97
3	18	23	38	43	58	63	78	83	98
2	19	22	39	42	59	62	79	82	99
1	20	21	40	41	60	61	80	81	100

1-100 Clockwise Spiral

1	2	3	4	5	6	7	8	9	10
36	37	38	39	40	41	42	43	44	11
35	64	65	66	67	68	69	70	45	12
34	63	84	85	86	87	88	71	46	13
33	62	83	96	97	98	89	72	47	14
32	61	82	95	100	99	90	73	48	15
31	60	81	94	93	92	91	74	49	16
30	59	80	79	78	77	76	75	50	17
29	58	57	56	55	54	53	52	51	18
28	27	26	25	24	23	22	21	20	19

1-100 Counterclockwise Spiral

1	36	35	34	33	32	31	30	29	28
2	37	64	63	62	61	60	59	58	27
3	38	65	84	83	82	81	80	57	26
4	39	66	85	96	95	94	79	56	25
5	40	67	86	97	100	93	78	55	24
6	41	68	87	98	99	92	77	54	23
7	42	69	88	89	90	91	76	53	22
8	43	70	71	72	73	74	75	52	21
9	44	45	46	47	48	49	50	51	20
10	11	12	13	14	15	16	17	18	19

1-100 Center Counterclockwise

100	99	98	97	96	95	94	93	92	91
65	64	63	62	61	60	59	58	57	90
66	37	36	35	34	33	32	31	56	89
67	38	17	16	15	14	13	30	55	88
68	39	18	5	4	3	12	29	54	87
69	40	19	6	1	2	11	28	53	86
70	41	20	7	8	9	10	27	52	85
71	42	21	22	23	24	25	26	51	84
72	43	44	45	46	47	48	49	50	83
73	74	75	76	77	78	79	80	81	82

1-100 Center Clock Spiral

73	74	75	76	77	78	79	80	81	82
72	43	44	45	46	47	48	49	50	83
71	42	21	22	23	24	25	26	51	84
70	41	20	7	8	9	10	27	52	85
69	40	19	6	1	2	11	28	53	86
68	39	18	5	4	3	12	29	54	87
67	38	17	16	15	14	13	30	55	88
66	37	36	35	34	33	32	31	56	89
65	64	63	62	61	60	59	58	57	90
100	99	98	97	96	95	94	93	92	91

1-100 Bottom Up Grid

91	92	93	94	95	96	97	98	99	100
81	82	83	84	85	86	87	88	89	90
71	72	73	74	75	76	77	78	79	80
61	62	63	64	65	66	67	68	69	70
51	52	53	54	55	56	57	58	59	60
41	42	43	44	45	46	47	48	49	50
31	32	33	34	35	36	37	38	39	40
21	22	23	24	25	26	27	28	29	30
11	12	13	14	15	16	17	18	19	20
1	2	3	4	5	6	7	8	9	10

Students can make their own spiral number grids using the table feature on a word processing or desktop publishing program.

Didax Educational Resources ~ www.didaxinc.com

Even Number Grid

2	4	6	8	10	12	14	16	18	20
22	24	26	28	30	32	34	36	38	40
42	44	46	48	50	52	54	56	58	60
62	64	66	68	70	72	74	76	78	80
82	84	86	88	90	92	94	96	98	100
102	104	106	108	110	112	114	116	118	120
122	124	126	128	130	132	134	136	138	140
142	144	146	148	150	152	154	156	158	160
162	164	166	168	170	172	174	176	178	180
182	184	186	188	190	192	194	196	198	200

Odd Number Grid

1	3	5	7	9	11	13	15	17	19
21	23	25	27	29	31	33	35	37	39
41	43	45	47	49	51	53	55	57	59
61	63	65	67	69	71	73	75	77	79
81	83	85	87	89	91	93	95	97	99
101	103	105	107	109	111	113	115	117	119
121	123	125	127	129	131	133	135	137	139
141	143	145	147	149	151	153	155	157	159
161	163	165	167	169	171	173	175	177	179
181	183	185	187	189	191	193	195	197	199

101-200 Number Grid

101	102	103	104	105	106	107	108	109	110
111	112	113	114	115	116	117	118	119	120
121	122	123	124	125	126	127	128	129	130
131	132	133	134	135	136	137	138	139	140
141	142	143	144	145	146	147	148	149	150
151	152	153	154	155	156	157	158	159	160
161	162	163	164	165	166	167	168	169	170
171	172	173	174	175	176	177	178	179	180
181	182	183	184	185	186	187	188	189	190
191	192	193	194	195	196	197	198	199	200

3 Number Grid

3	6	9	12	15	18	21	24	27	30
33	36	39	42	45	48	51	54	57	60
63	66	69	72	75	78	81	84	87	90
93	96	99	102	105	108	111	114	117	120
123	126	129	132	135	138	141	144	147	150
153	156	159	162	165	168	171	174	177	180
183	186	189	192	195	198	201	204	207	210
213	216	219	222	225	228	231	234	237	240
243	246	249	252	255	258	261	264	267	270
273	276	279	282	285	288	291	294	297	300

5-500 Number Grid

5	10	15	20	25	30	35	40	45	50
55	60	65	70	75	80	85	90	95	100
105	110	115	120	125	130	135	140	145	150
155	160	165	170	175	180	185	190	195	200
205	210	215	220	225	230	235	240	245	250
255	260	265	270	275	280	285	290	295	300
305	310	315	320	325	330	335	340	345	350
355	360	365	370	375	380	385	390	395	400
405	410	415	420	425	430	435	440	445	450
455	460	465	470	475	480	485	490	495	500

0.1-10.0 Number Grid

0.1	0.2	0.3	0.4	0.5	0.6	0.7	0.8	0.9	1.0
1.1	1.2	1.3	1.4	1.5	1.6	1.7	1.8	1.9	2.0
2.1	2.2	2.3	2.4	2.5	2.6	2.7	2.8	2.9	3.0
3.1	3.2	3.3	3.4	3.5	3.6	3.7	3.8	3.9	4.0
4.1	4.2	4.3	4.4	4.5	4.6	4.7	4.8	4.9	5.0
5.1	5.2	5.3	5.4	5.5	5.6	5.7	5.8	5.9	6.0
6.1	6.2	6.3	6.4	6.5	6.6	6.7	6.8	6.9	7.0
7.1	7.2	7.3	7.4	7.5	7.6	7.7	7.8	7.9	8.0
8.1	8.2	8.3	8.4	8.5	8.6	8.7	8.8	8.9	9.0
9.1	9.2	9.3	9.4	9.5	9.6	9.7	9.8	9.9	10.0

Number Grids

9x9 Grid

1	2	3	4	5	6	7	8	9
10	11	12	13	14	15	16	17	18
19	20	21	22	23	24	25	26	27
28	29	30	31	32	33	34	35	36
37	38	39	40	41	42	43	44	45
46	47	48	49	50	51	52	53	54
55	56	57	58	59	60	61	62	63
64	65	66	67	68	69	70	71	72
73	74	75	76	77	78	79	80	81

8x8 Grid

1	2	3	4	5	6	7	8
9	10	11	12	13	14	15	16
17	18	19	20	21	22	23	24
25	26	27	28	29	30	31	32
33	34	35	36	37	38	39	40
41	42	43	44	45	46	47	48
49	50	51	52	53	54	55	56
57	58	59	60	61	62	63	64

7x7 Grid

1	2	3	4	5	6	7
8	9	10	11	12	13	14
15	16	17	18	19	20	21
22	23	24	25	26	27	28
29	30	31	32	33	34	35
36	37	38	39	40	41	42
43	44	45	46	47	48	49

6x6 Grid

1	2	3	4	5	6
7	8	9	10	11	12
13	14	15	16	17	18
19	20	21	22	23	24
25	26	27	28	29	30
31	32	33	34	35	36

5x5 Grid

1	2	3	4	5
6	7	8	9	10
11	12	13	14	15
16	17	18	19	20
21	22	23	24	25

4x4 Grid

1	2	3	4
5	6	7	8
9	10	11	12
13	14	15	16

Didax Educational Resources ~ www.didaxinc.com

11x11 Number Grid

1	2	3	4	5	6	7	8	9	10	11
12	13	14	15	16	17	18	19	20	21	22
23	24	25	26	27	28	29	30	31	32	33
34	35	36	37	38	39	40	41	42	43	44
45	46	47	48	49	50	51	52	53	54	55
56	57	58	59	60	61	62	63	64	65	66
67	68	69	70	71	72	73	74	75	76	77
78	79	80	81	82	83	84	85	86	87	88
89	90	91	92	93	94	95	96	97	98	99
100	101	102	103	104	105	106	107	108	109	110
111	112	113	114	115	116	117	118	119	120	121

12x12 Number Grid

1	2	3	4	5	6	7	8	9	10	11	12
13	14	15	16	17	18	19	20	21	22	23	24
25	26	27	28	29	30	31	32	33	34	35	36
37	38	39	40	41	42	43	44	45	46	47	48
49	50	51	52	53	54	55	56	57	58	59	60
61	62	63	64	65	66	67	68	69	70	71	72
73	74	75	76	77	78	79	80	81	82	83	84
85	86	87	88	89	90	91	92	93	94	95	96
97	98	99	100	101	102	103	104	105	106	107	108
109	110	111	112	113	114	115	116	117	118	119	120
121	122	123	124	125	126	127	128	129	130	131	132
133	134	135	136	137	138	139	140	141	142	143	144

13x13 Number Grid

1	2	3	4	5	6	7	8	9	10	11	12	13
14	15	16	17	18	19	20	21	22	23	24	25	26
27	28	29	30	31	32	33	34	35	36	37	38	39
40	41	42	43	44	45	46	47	48	49	50	51	52
53	54	55	56	57	58	59	60	61	62	63	64	65
66	67	68	69	70	71	72	73	74	75	76	77	78
79	80	81	82	83	84	85	86	87	88	89	90	91
92	93	94	95	96	97	98	99	100	101	102	103	104
105	106	107	108	109	110	111	112	113	114	115	116	117
118	119	120	121	122	123	124	125	126	127	128	129	130
131	132	133	134	135	136	137	138	139	140	141	142	143
144	145	146	147	148	149	150	151	152	153	154	155	156
157	158	159	160	161	162	163	164	165	166	167	168	169

14x14 Number Grid

1	2	3	4	5	6	7	8	9	10	11	12	13	14
15	16	17	18	19	20	21	22	23	24	25	26	27	28
29	30	31	32	33	34	35	36	37	38	39	40	41	42
43	44	45	46	47	48	49	50	51	52	53	54	55	56
57	58	59	60	61	62	63	64	65	66	67	68	69	70
71	72	73	74	75	76	77	78	79	80	81	82	83	84
85	86	87	88	89	90	91	92	93	94	95	96	97	98
99	100	101	102	103	104	105	106	107	108	109	110	111	112
113	114	115	116	117	118	119	120	121	122	123	124	125	126
127	128	129	130	131	132	133	134	135	136	137	138	139	140
141	142	143	144	145	146	147	148	149	150	151	152	153	154
155	156	157	158	159	160	161	162	163	164	165	166	167	168
169	170	171	172	173	174	175	176	177	178	179	180	181	182
183	184	185	186	187	188	189	190	191	192	193	194	195	196

15x15 Number Grid

1	2	3	4	5	6	7	8	9	10	11	12	13	14	15
16	17	18	19	20	21	22	23	24	25	26	27	28	29	30
31	32	33	34	35	36	37	38	39	40	41	42	43	44	45
46	47	48	49	50	51	52	53	54	55	56	57	58	59	60
61	62	63	64	65	66	67	68	69	70	71	72	73	74	75
76	77	78	79	80	81	82	83	84	85	86	87	88	89	90
91	92	93	94	95	96	97	98	99	100	101	102	103	104	105
106	107	108	109	110	111	112	113	114	115	116	117	118	119	120
121	122	123	124	125	126	127	128	129	130	131	132	133	134	135
136	137	138	139	140	141	142	143	144	145	146	147	148	149	150
151	152	153	154	155	156	157	158	159	160	161	162	163	164	165
166	167	168	169	170	171	172	173	174	175	176	177	178	179	180
181	182	183	184	185	186	187	188	189	190	191	192	193	194	195
196	197	198	199	200	201	202	203	204	205	206	207	208	209	210
211	212	213	214	215	216	217	218	219	220	221	222	223	224	225

16x16 Number Grid

1	2	3	4	5	6	7	8	9	10	11	12	13	14	15	16
17	18	19	20	21	22	23	24	25	26	27	28	29	30	31	32
33	34	35	36	37	38	39	40	41	42	43	44	45	46	47	48
49	50	51	52	53	54	55	56	57	58	59	60	61	62	63	64
65	66	67	68	69	70	71	72	73	74	75	76	77	78	79	80
81	82	83	84	85	86	87	88	89	90	91	92	93	94	95	96
97	98	99	100	101	102	103	104	105	106	107	108	109	110	111	112
113	114	115	116	117	118	119	120	121	122	123	124	125	126	127	128
129	130	131	132	133	134	135	136	137	138	139	140	141	142	143	144
145	146	147	148	149	150	151	152	153	154	155	156	157	158	159	160
161	162	163	164	165	166	167	168	169	170	171	172	173	174	175	176
177	178	179	180	181	182	183	184	185	186	187	188	189	190	191	192
193	194	195	196	197	198	199	200	201	202	203	204	205	206	207	208
209	210	211	212	213	214	215	216	217	218	219	220	221	222	223	224
225	226	227	228	229	230	231	232	233	234	235	236	237	238	239	240
241	242	243	244	245	246	247	248	249	250	251	252	253	254	255	256

Number Grids

17x17 Number Grid

1	2	3	4	5	6	7	8	9	10	11	12	13	14	15	16	17
18	19	20	21	22	23	24	25	26	27	28	29	30	31	32	33	34
35	36	37	38	39	40	41	42	43	44	45	46	47	48	49	50	51
52	53	54	55	56	57	58	59	60	61	62	63	64	65	66	67	68
69	70	71	72	73	74	75	76	77	78	79	80	81	82	83	84	85
86	87	88	89	90	91	92	93	94	95	96	97	98	99	100	101	102
103	104	105	106	107	108	109	110	111	112	113	114	115	116	117	118	119
120	121	122	123	124	125	126	127	128	129	130	131	132	133	134	135	136
137	138	139	140	141	142	143	144	145	146	147	148	149	150	151	152	153
154	155	156	157	158	159	160	161	162	163	164	165	166	167	168	169	170
171	172	173	174	175	176	177	178	179	180	181	182	183	184	185	186	187
188	189	190	191	192	193	194	195	196	197	198	199	200	201	202	203	204
205	206	207	208	209	210	211	212	213	214	215	216	217	218	219	220	221
222	223	224	225	226	227	228	229	230	231	232	233	234	235	236	237	238
239	240	241	242	243	244	245	246	247	248	249	250	251	252	253	254	255
256	257	258	259	260	261	262	263	264	265	266	267	268	269	270	271	272
273	274	275	276	277	278	279	280	281	282	283	284	285	286	287	288	289

Didax Educational Resources ~ www.didaxinc.com

18x18 Number Grid

1	2	3	4	5	6	7	8	9	10	11	12	13	14	15	16	17	18
19	20	21	22	23	24	25	26	27	28	29	30	31	32	33	34	35	36
37	38	39	40	41	42	43	44	45	46	47	48	49	50	51	52	53	54
55	56	57	58	59	60	61	62	63	64	65	66	67	68	69	70	71	72
73	74	75	76	77	78	79	80	81	82	83	84	85	86	87	88	89	90
91	92	93	94	95	96	97	98	99	100	101	102	103	104	105	106	107	108
109	110	111	112	113	114	115	116	117	118	119	120	121	122	123	124	125	126
127	128	129	130	131	132	133	134	135	136	137	138	139	140	141	142	143	144
145	146	147	148	149	150	151	152	153	154	155	156	157	158	159	160	161	162
163	164	165	166	167	168	169	170	171	172	173	174	175	176	177	178	179	180
181	182	183	184	185	186	187	188	189	190	191	192	193	194	195	196	197	198
199	200	201	202	203	204	205	206	207	208	209	210	211	212	213	214	215	216
217	218	219	220	221	222	223	224	225	226	227	228	229	230	231	232	233	234
235	236	237	238	239	240	241	242	243	244	245	246	247	248	249	250	251	252
253	254	255	256	257	258	259	260	261	262	263	264	265	266	267	268	269	270
271	272	273	274	275	276	277	278	279	280	281	282	283	284	285	286	287	288
289	290	291	292	293	294	295	296	297	298	299	300	301	302	303	304	305	306
307	308	309	310	311	312	313	314	315	316	317	318	319	320	321	322	323	324

19x19 Number Grid

1	2	3	4	5	6	7	8	9	10	11	12	13	14	15	16	17	18	19
20	21	22	23	24	25	26	27	28	29	30	31	32	33	34	35	36	37	38
39	40	41	42	43	44	45	46	47	48	49	50	51	52	53	54	55	56	57
58	59	60	61	62	63	64	65	66	67	68	69	70	71	72	73	74	75	76
77	78	79	80	81	82	83	84	85	86	87	88	89	90	91	92	93	94	95
96	97	98	99	100	101	102	103	104	105	106	107	108	109	110	111	112	113	114
115	116	117	118	119	120	121	122	123	124	125	126	127	128	129	130	131	132	133
134	135	136	137	138	139	140	141	142	143	144	145	146	147	148	149	150	151	152
153	154	155	156	157	158	159	160	161	162	163	164	165	166	167	168	169	170	171
172	173	174	175	176	177	178	179	180	181	182	183	184	185	186	187	188	189	190
191	192	193	194	195	196	197	198	199	200	201	202	203	204	205	206	207	208	209
210	211	212	213	214	215	216	217	218	219	220	221	222	223	224	225	226	227	228
229	230	231	232	233	234	235	236	237	238	239	240	241	242	243	244	245	246	247
248	249	250	251	252	253	254	255	256	257	258	259	260	261	262	263	264	265	266
267	268	269	270	271	272	273	274	275	276	277	278	279	280	281	282	283	284	285
286	287	288	289	290	291	292	293	294	295	296	297	298	299	300	301	302	303	304
305	306	307	308	309	310	311	312	313	314	315	316	317	318	319	320	321	322	323
324	325	326	327	328	329	330	331	332	333	334	335	336	337	338	339	340	341	342
343	344	345	346	347	348	349	350	351	352	353	354	355	356	357	358	359	360	361

20x20 Number Grid

1	2	3	4	5	6	7	8	9	10	11	12	13	14	15	16	17	18	19	20
21	22	23	24	25	26	27	28	29	30	31	32	33	34	35	36	37	38	39	40
41	42	43	44	45	46	47	48	49	50	51	52	53	54	55	56	57	58	59	60
61	62	63	64	65	66	67	68	69	70	71	72	73	74	75	76	77	78	79	80
81	82	83	84	85	86	87	88	89	90	91	92	93	94	95	96	97	98	99	100
101	102	103	104	105	106	107	108	109	110	111	112	113	114	115	116	117	118	119	120
121	122	123	124	125	126	127	128	129	130	131	132	133	134	135	136	137	138	139	140
141	142	143	144	145	146	147	148	149	150	151	152	153	154	155	156	157	158	159	160
161	162	163	164	165	166	167	168	169	170	171	172	173	174	175	176	177	178	179	180
181	182	183	184	185	186	187	188	189	190	191	192	193	194	195	196	197	198	199	200
201	202	203	204	205	206	207	208	209	210	211	212	213	214	215	216	217	218	219	220
221	222	223	224	225	226	227	228	229	230	231	232	233	234	235	236	237	238	239	240
241	242	243	244	245	246	247	248	249	250	251	252	253	254	255	256	257	258	259	260
261	262	263	264	265	266	267	268	269	270	271	272	273	274	275	276	277	278	279	280
281	282	283	284	285	286	287	288	289	290	291	292	293	294	295	296	297	298	299	300
301	302	303	304	305	306	307	308	309	310	311	312	313	314	315	316	317	318	319	320
321	322	323	324	325	326	327	328	329	330	331	332	333	334	335	336	337	338	339	340
341	342	343	344	345	346	347	348	349	350	351	352	353	354	355	356	357	358	359	360
361	362	363	364	365	366	367	368	369	370	371	372	373	374	375	376	377	378	379	380
381	382	383	384	385	386	387	388	389	390	391	392	393	394	395	396	397	398	399	400

0-99 Number Grid

0	1	2	3	4	5	6	7	8	9
10	11	12	13	14	15	16	17	18	19
20	21	22	23	24	25	26	27	28	29
30	31	32	33	34	35	36	37	38	39
40	41	42	43	44	45	46	47	48	49
50	51	52	53	54	55	56	57	58	59
60	61	62	63	64	65	66	67	68	69
70	71	72	73	74	75	76	77	78	79
80	81	82	83	84	85	86	87	88	89
90	91	92	93	94	95	96	97	98	99

0-99 Number Grid

0	1	2	3	4	5	6	7	8	9
10	11	12	13	14	15	16	17	18	19
20	21	22	23	24	25	26	27	28	29
30	31	32	33	34	35	36	37	38	39
40	41	42	43	44	45	46	47	48	49
50	51	52	53	54	55	56	57	58	59
60	61	62	63	64	65	66	67	68	69
70	71	72	73	74	75	76	77	78	79
80	81	82	83	84	85	86	87	88	89
90	91	92	93	94	95	96	97	98	99

0-99 Number Grid

0	1	2	3	4	5	6	7	8	9
10	11	12	13	14	15	16	17	18	19
20	21	22	23	24	25	26	27	28	29
30	31	32	33	34	35	36	37	38	39
40	41	42	43	44	45	46	47	48	49
50	51	52	53	54	55	56	57	58	59
60	61	62	63	64	65	66	67	68	69
70	71	72	73	74	75	76	77	78	79
80	81	82	83	84	85	86	87	88	89
90	91	92	93	94	95	96	97	98	99

0-99 Number Grid

0	1	2	3	4	5	6	7	8	9
10	11	12	13	14	15	16	17	18	19
20	21	22	23	24	25	26	27	28	29
30	31	32	33	34	35	36	37	38	39
40	41	42	43	44	45	46	47	48	49
50	51	52	53	54	55	56	57	58	59
60	61	62	63	64	65	66	67	68	69
70	71	72	73	74	75	76	77	78	79
80	81	82	83	84	85	86	87	88	89
90	91	92	93	94	95	96	97	98	99

0-99 Number Grid

0	1	2	3	4	5	6	7	8	9
10	11	12	13	14	15	16	17	18	19
20	21	22	23	24	25	26	27	28	29
30	31	32	33	34	35	36	37	38	39
40	41	42	43	44	45	46	47	48	49
50	51	52	53	54	55	56	57	58	59
60	61	62	63	64	65	66	67	68	69
70	71	72	73	74	75	76	77	78	79
80	81	82	83	84	85	86	87	88	89
90	91	92	93	94	95	96	97	98	99

0-99 Number Grid

0	1	2	3	4	5	6	7	8	9
10	11	12	13	14	15	16	17	18	19
20	21	22	23	24	25	26	27	28	29
30	31	32	33	34	35	36	37	38	39
40	41	42	43	44	45	46	47	48	49
50	51	52	53	54	55	56	57	58	59
60	61	62	63	64	65	66	67	68	69
70	71	72	73	74	75	76	77	78	79
80	81	82	83	84	85	86	87	88	89
90	91	92	93	94	95	96	97	98	99

1-100 Number Grid

1	2	3	4	5	6	7	8	9	10
11	12	13	14	15	16	17	18	19	20
21	22	23	24	25	26	27	28	29	30
31	32	33	34	35	36	37	38	39	40
41	42	43	44	45	46	47	48	49	50
51	52	53	54	55	56	57	58	59	60
61	62	63	64	65	66	67	68	69	70
71	72	73	74	75	76	77	78	79	80
81	82	83	84	85	86	87	88	89	90
91	92	93	94	95	96	97	98	99	100

1-100 Number Grid

1	2	3	4	5	6	7	8	9	10
11	12	13	14	15	16	17	18	19	20
21	22	23	24	25	26	27	28	29	30
31	32	33	34	35	36	37	38	39	40
41	42	43	44	45	46	47	48	49	50
51	52	53	54	55	56	57	58	59	60
61	62	63	64	65	66	67	68	69	70
71	72	73	74	75	76	77	78	79	80
81	82	83	84	85	86	87	88	89	90
91	92	93	94	95	96	97	98	99	100

1-100 Number Grid

1	2	3	4	5	6	7	8	9	10
11	12	13	14	15	16	17	18	19	20
21	22	23	24	25	26	27	28	29	30
31	32	33	34	35	36	37	38	39	40
41	42	43	44	45	46	47	48	49	50
51	52	53	54	55	56	57	58	59	60
61	62	63	64	65	66	67	68	69	70
71	72	73	74	75	76	77	78	79	80
81	82	83	84	85	86	87	88	89	90
91	92	93	94	95	96	97	98	99	100

1-100 Number Grid

1	2	3	4	5	6	7	8	9	10
11	12	13	14	15	16	17	18	19	20
21	22	23	24	25	26	27	28	29	30
31	32	33	34	35	36	37	38	39	40
41	42	43	44	45	46	47	48	49	50
51	52	53	54	55	56	57	58	59	60
61	62	63	64	65	66	67	68	69	70
71	72	73	74	75	76	77	78	79	80
81	82	83	84	85	86	87	88	89	90
91	92	93	94	95	96	97	98	99	100

1-100 Number Grid

1	2	3	4	5	6	7	8	9	10
11	12	13	14	15	16	17	18	19	20
21	22	23	24	25	26	27	28	29	30
31	32	33	34	35	36	37	38	39	40
41	42	43	44	45	46	47	48	49	50
51	52	53	54	55	56	57	58	59	60
61	62	63	64	65	66	67	68	69	70
71	72	73	74	75	76	77	78	79	80
81	82	83	84	85	86	87	88	89	90
91	92	93	94	95	96	97	98	99	100

1-100 Number Grid

1	2	3	4	5	6	7	8	9	10
11	12	13	14	15	16	17	18	19	20
21	22	23	24	25	26	27	28	29	30
31	32	33	34	35	36	37	38	39	40
41	42	43	44	45	46	47	48	49	50
51	52	53	54	55	56	57	58	59	60
61	62	63	64	65	66	67	68	69	70
71	72	73	74	75	76	77	78	79	80
81	82	83	84	85	86	87	88	89	90
91	92	93	94	95	96	97	98	99	100

Number Grids

100 Square Grid

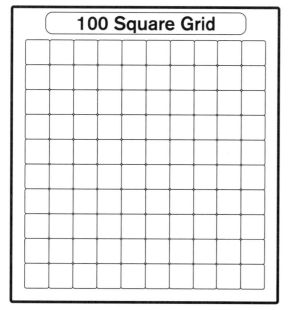

100 Square Grid

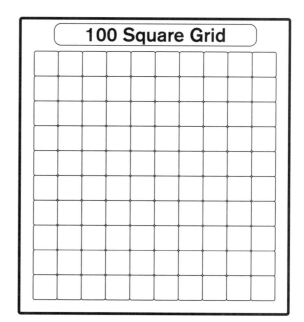

100 Square Grid

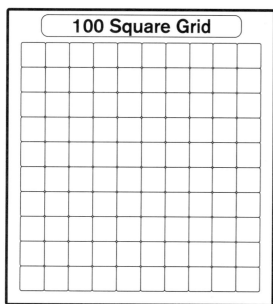

100 Square Grid

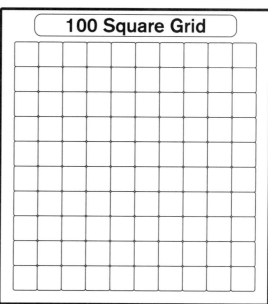

100 Square Grid

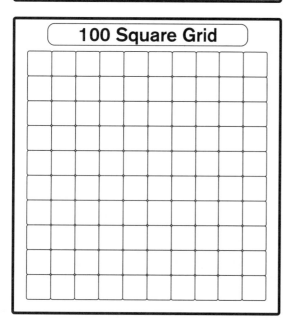

100 Square Grid

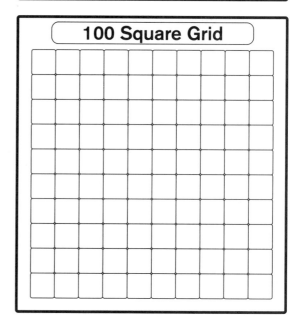

1-99 Number Grid

	1	2	3	4	5	6	7	8	9
10	11	12	13	14	15	16	17	18	19
20	21	22	23	24	25	26	27	28	29
30	31	32	33	34	35	36	37	38	39
40	41	42	43	44	45	46	47	48	49
50	51	52	53	54	55	56	57	58	59
60	61	62	63	64	65	66	67	68	69
70	71	72	73	74	75	76	77	78	79
80	81	82	83	84	85	86	87	88	89
90	91	92	93	94	95	96	97	98	99

1-99 Number Grid

	1	2	3	4	5	6	7	8	9
10	11	12	13	14	15	16	17	18	19
20	21	22	23	24	25	26	27	28	29
30	31	32	33	34	35	36	37	38	39
40	41	42	43	44	45	46	47	48	49
50	51	52	53	54	55	56	57	58	59
60	61	62	63	64	65	66	67	68	69
70	71	72	73	74	75	76	77	78	79
80	81	82	83	84	85	86	87	88	89
90	91	92	93	94	95	96	97	98	99

1-99 Number Grid

	1	2	3	4	5	6	7	8	9
10	11	12	13	14	15	16	17	18	19
20	21	22	23	24	25	26	27	28	29
30	31	32	33	34	35	36	37	38	39
40	41	42	43	44	45	46	47	48	49
50	51	52	53	54	55	56	57	58	59
60	61	62	63	64	65	66	67	68	69
70	71	72	73	74	75	76	77	78	79
80	81	82	83	84	85	86	87	88	89
90	91	92	93	94	95	96	97	98	99

1-99 Number Grid

	1	2	3	4	5	6	7	8	9
10	11	12	13	14	15	16	17	18	19
20	21	22	23	24	25	26	27	28	29
30	31	32	33	34	35	36	37	38	39
40	41	42	43	44	45	46	47	48	49
50	51	52	53	54	55	56	57	58	59
60	61	62	63	64	65	66	67	68	69
70	71	72	73	74	75	76	77	78	79
80	81	82	83	84	85	86	87	88	89
90	91	92	93	94	95	96	97	98	99

1-99 Number Grid

	1	2	3	4	5	6	7	8	9
10	11	12	13	14	15	16	17	18	19
20	21	22	23	24	25	26	27	28	29
30	31	32	33	34	35	36	37	38	39
40	41	42	43	44	45	46	47	48	49
50	51	52	53	54	55	56	57	58	59
60	61	62	63	64	65	66	67	68	69
70	71	72	73	74	75	76	77	78	79
80	81	82	83	84	85	86	87	88	89
90	91	92	93	94	95	96	97	98	99

1-99 Number Grid

	1	2	3	4	5	6	7	8	9
10	11	12	13	14	15	16	17	18	19
20	21	22	23	24	25	26	27	28	29
30	31	32	33	34	35	36	37	38	39
40	41	42	43	44	45	46	47	48	49
50	51	52	53	54	55	56	57	58	59
60	61	62	63	64	65	66	67	68	69
70	71	72	73	74	75	76	77	78	79
80	81	82	83	84	85	86	87	88	89
90	91	92	93	94	95	96	97	98	99

Number Grids

1-100 Number Grid

1	2	3	4	5	6	7	8	9	10
11	12	13	14	15	16	17	18	19	20
21	22	23	24	25	26	27	28	29	30
31	32	33	34	35	36	37	38	39	40
41	42	43	44	45	46	47	48	49	50
51	52	53	54	55	56	57	58	59	60
61	62	63	64	65	66	67	68	69	70
71	72	73	74	75	76	77	78	79	80
81	82	83	84	85	86	87	88	89	90
91	92	93	94	95	96	97	98	99	100

This larger 1-100 number grid can be used to assist students in checking and making the jigsaw and missing number activities on pages 48 - 50 in Section 3.

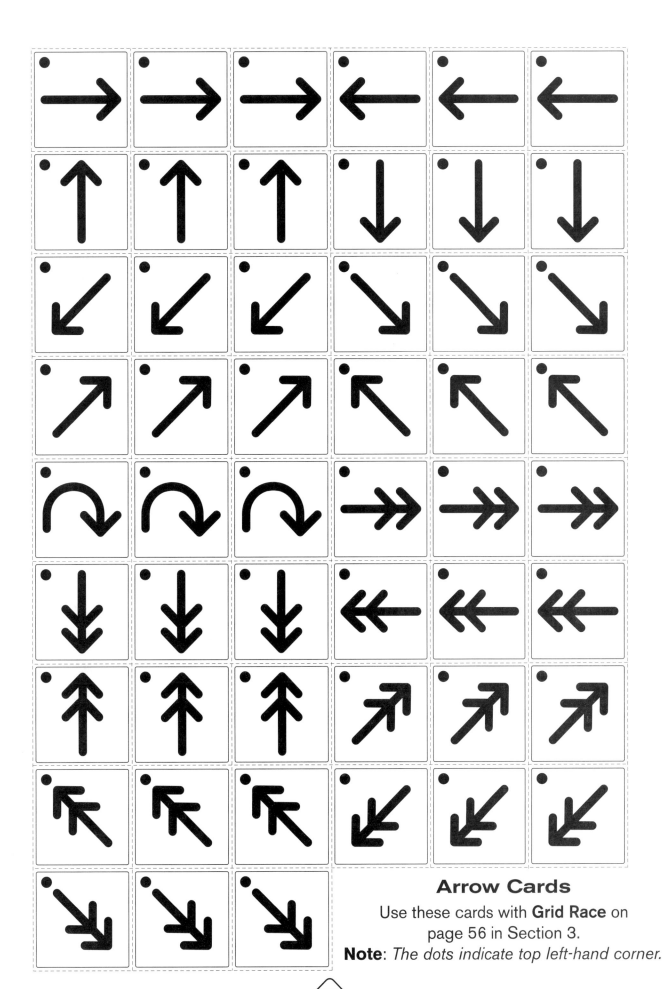

Arrow Cards

Use these cards with **Grid Race** on page 56 in Section 3.

Note: *The dots indicate top left-hand corner.*

Activity and Game Notes and Answers
pages 18-27

In this section, you will find detailed background information about each game and activity in Section 3.

The purpose of each game and activity is explained, along with teaching points and ideas you can use to assist students' understanding.

Answers are also provided where necessary.

s
e
c
t
i
o
n

2

Activity and Game Notes and Answers

Cut and Paste

Page 43

This exercise is designed to help students see the relationship between number tracks and rolls and the more compact number grid. Many students are surprised at the length of the number track. Making a number roll also helps students become familiar with number grids. Both grids are designed to be cut and glued.

Count 'n Color

Page 44

The constant counting function of the calculator may be used in conjunction with a number grid to support students' early counting. Most calculators can be set to count constant amounts by pressing + constant = = = . For example, to count in fives you would press + 5 = = = .

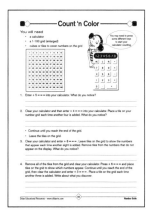

Note: Different model calculators may require the use of a different sequence e.g. 5 + + = . Experiment with the models available in the classroom. To vary the starting number use the following sequence: START NUMBER + STEP NUMBER = = = . For example, to start at two and count by fives you would press 2 + 5 = = = .

1. + 5 = = = produces a 5, 10, 15, 20... counting sequence.

2. + 4 = = = produces the multiples of 4 (4, 8, 12, 16...) Students should notice that a cell is marked in the same column, but every other now under 4, 8, 12 and 20.

3. Students should notice that they have to remove every second cube because every second multiple of 4 is also a multiple of 8 (4, **8**, 12, **16**, 20, **24**, 28, **32**...).

4. Pressing + 6 = = = will produce the multiples of 6 (6, 12, 18, 24...). Students should notice that + 3 = = = produces the multiples of 3 (3, **6**, 9, **12**...), of which every second multiple is also a multiple of 6. Students will have to place extra cubes on the grid at 3, 9,15, 21, 27...

Hidden Numbers

Page 45

These games help students to become more familiar with the configuration of a grid. To determine which number or numbers are hidden, students need to make use of their number sense.

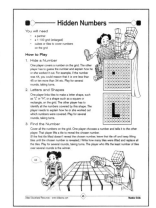

For example, to work out that 44 is the hidden number a student might reason:
the missing number is one more than 43, one less than 45, ten more than 34, ten less than 54, or that the hidden number is in the fourth column and the fourth row; therefore, it must be 44.

(Different grids can be used.)

Missing Numbers

Page 46

This activity develops similar thinking to **Hidden Numbers**. The difference is that less numbers are provided as clues, so students have to use the given numbers as the starting point for determining the appropriate cells for the missing numbers.

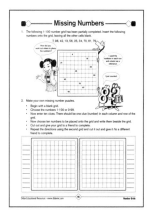

Number Grid Games

Page 47

Familiarity with the layout of various number grids may be developed by using the games outlined on this page. An understanding of place value along with the use of some strategy are required to play the game successfully. The game may be played on either a 0-99 or 1-100 grid.

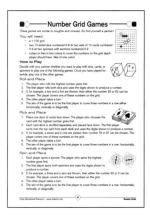

With a few adjustments, any grid configuration can be used.

Jigsaws – 1

Page 48

The jigsaw activity helps students focus on various relationships within the grid in order to match pieces. For example, horizontal pieces fit together on a one more, one less basis; vertical pieces fit together on a ten more, ten less basis. Diagonal pieces rely on an eleven more, eleven less relationship.

When students make their own jigsaws, the patterns become more obvious.

Note: See page 26 for an enlarged 1-100 grid to assist with this activity.

More Missing Numbers

Page 49

In order to fill in the missing numbers on the grid pieces students need to be able to add 1, 10 and 11 and subtract 1, 10 and 11. An understanding of the patterns in the grid will help the students complete the puzzle. When students make up their own puzzles, the patterns become more obvious.

Note: Answers can be checked using the 1-100 grid on page 26.

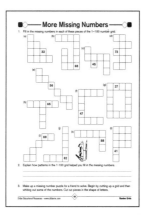

Jigsaws – 2

Page 50

Once students have mastered reassembling 1-100 or 0-99 grids, less familiar grids can be used. Blocking out numbers provides a further degree of difficulty.

Note: See page 26 for an enlarged 1-100 grid to assist with this activity.

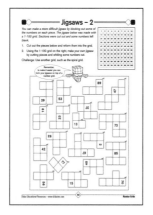

Placing for Points

Page 51

The aim of the game is to place a numbered piece on the correct place on a grid. These can be cut out from the grid on page 26.

All the number prompts are removed in this game and students must rely on their knowledge of the various patterns in the grid to place the pieces correctly.

Placing 100 numbers on a grid can become a little tedious, so the following variations are suggested. (Use the appropriate set of numbers for each activity.)

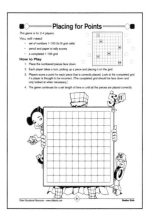

- Use a smaller grid, for example; 1-20 for younger students, 1-50, 27-77, 50-100, etc.

- Use a grid with the odd or even numbers blocked out. Students then pick up either odd or even numbers and place them on the grid.

- Use a multiple grid. For example, the multiples of seven may be blocked out on the grid. When a student picks up and places a multiple of seven correctly he or she receives a bonus point. The multiples of four and seven could be blocked out. A player could be awarded a bonus point when a multiple of four or a multiple of seven is placed. Two bonus points could be awarded for placing a multiple of four and seven (28, 56, 84, etc.)

Number Spirals

Page 52

Students will observe many patterns when creating their own number grids (refer to page 9). The spiral grids are different from the previous number grids and will challenge many students' thinking.

Note: See page 12 for completed grids.

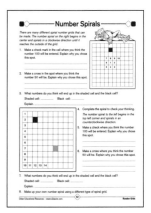

Activity and Game Notes and Answers

All the Right Moves – 1-3
Pages 53 - 55

This activity reinforces many of the addition and subtraction ideas encountered earlier in activities such as **Hide a Number** and **Jigsaws**. The use of symbols (the various arrows) to represent adding and subtracting 1, 9, 10 and 11 is an introduction to the use of symbols in algebra.

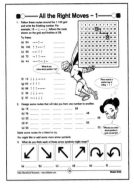

The relationships between the numbers on the grid may be established by following a routine similar to the one outlined below.

*   Ask each student to place a marker on any number in the middle of the grid. Have the student observe the following:

    – What number is directly above your number?

    – What number is directly below your number?

    – What number is to the right of your number?

    – What number is to the left of your number?

*   Start with another number and repeat the sequence. Ask the students what they notice. Many will soon notice that the numbers are either ten less, ten more, one more or one less than the starting number.

*   Next, ask the students to consider the numbers two cells above or below the starting number. What about numbers two cells to the right or left of the starting number?

*   Introduce the use of arrows to describe moves on the grid.

    For example; **q** up one cell   **Q** down one cell

    R right one cell   **r** left one cell

*   Describe a route from one number to another. For example 15 **R RQ**. This route would finish at 27. Ask the students to produce routes for their friends to follow.

Describing the route from one number to another can be quite demanding. As the activity progresses, students should be encouraged to look for short-cut systems for combining several arrows. For example ↑↑↑ might become ⬆.

All the Right Moves – 1

Answers

1. (a) 47   (b) 68   (c) 37   (d) 55   (e) 44

    (f) 46   (g) 31   (h) 53   (i) 67   (j) 99

2. Answers may vary.

3. (a) +9   (b) +11 (c) −11 (d) −9 (e) +2 (f) −2

All the Right Moves – 2

In this set of arrows, some new symbols are introduced. Rather than tell the students what they might mean, give them a chance to suggest what they might represent. ↓ is simply a shorthand way of writing ↓↓. It is important students come to the realization that symbols are a powerful means of describing something in a clear and concise manner. The symbols ⤳ and → also provide an opportunity for discussion. Both symbols represent the same action–moving two cells to the right. At first students may seem a little puzzled by this, but they shouldn't be. For example, the symbols ÷ and) are both used to describe division.

You may wish to talk about developing some more symbols. For example, ⤳̄⁴ represents a move four cells to the right.

Answers

1. subtract 20
2. (a) −2 (b) +18 (c) −22 (d) +20 (e) +22 (f) −18
3. (a) −5 (b) +45 (c) −55 (d) +50 (e) +55 (f) −45
4. Answers will vary.
5. The value of each arrow changes by a factor that matches the multiples shown on the grid. For example, on a standard 1-100 grid numbered from the top down → means +1. On a 2-200 grid it means +2, on a 3–300 grid it means +3... on a 6–600 grid it would mean +6. The arrow ↗ which would mean −9 on the 1-100 grid would mean −18 on the 2-200 grid, −27 on a 3-300 grid, −45 on a 5-500 grid etc.

A further investigation involves considering what happens if the grid were numbered from the bottom up. (Refer to the Bottom Up Grid on page 12.) Now → still means +1 but ↓ means −10 and ↑ means +10, ↘ means +9 and means +11. Note which arrows change and which stay the same.

All the Right Moves – 3

Answers

1. (a) +5 (b) −4 (c) +4

2. Answers will vary depending on the grid size chosen.

3. The grid size determines the value of the ↑ and ↓ arrows. For example, on a 4x4 number grid ↑ and ↓ have a value of four and the direction of the arrow determines whether addition or subtraction is used. ↑ represents subtraction and ↓ addition.

    On a 6x6 number grid ↑ and ↓ would have a value of six. On a 7x7 grid, the value is seven and so on. The → and ← arrows still represent +1 and −1 regardless of the grid size. As a guide:

    ↑ – grid size, ↓ + grid size, → +1, ← −1

    ↖ – one more than grid size

    ↗ – one less than grid size

    ↙ + one less than grid size

    ↘ + one more than grid size

Grid Race
Page 56

This is an extension of the ideas found in All The Right Moves. Students will need a set of arrow cards found on page 27. The use of a die or spinner adds an element of chance. Refer to the student page for other variations. You can provide students with simple addition cards only, or extra arrow cards where they choose a number to add or subtract.

Think of a Number
Page 57

This game is popular in many classes but there are always some students who become a little lost while playing it. Shading the grid after each question is asked helps in two ways:

1. It helps students to keep track of what is going on, preventing them from becoming lost.

2. It highlights which questions are best to ask narrowing the field of possible numbers to choose from.

After playing the game a few times, students can be challenged to find the missing number using seven questions or less.

Eratosthenes' Sieve
Page 58

Answers

2. Prime numbers are found in the first, third, seventh and ninth columns.

3. When marking the prime numbers on a six-column grid, students should notice that after the first row the prime numbers are found in the first or fifth column.

Marking Multiples – 1 and 2
Pages 59 - 60

Answers

Multiples of 2

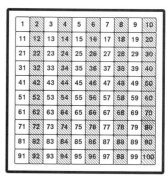

Multiples of 3

Multiples of 4

Multiples of 5

Marking Multiples – 1-2
Pages 59 - 60

Multiples of 6

Multiples of 10

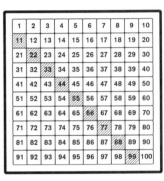

Multiples of 7

Multiples of 11

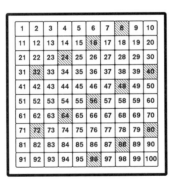

Multiples of 8

Multiples of 12

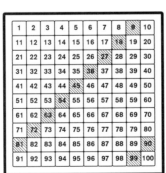

Further ideas for looking at patterns within multiple patterns:

- Consider patterns of digits that add to 11, or another number.

- Explore multiple patterns on other grids. For example, a standard grid compared to a spiral grid.

- Consider other patterns such as square numbers, triangular numbers and Fibonacci numbers.

- Teachers may introduce the notion of vectors when describing patterns on a number grid. For example, if r=right, l=left, u=up and d=down one cell respectively, then patterns may be described using a combination of moves to form an expression. The expression 2r + d could be used to describe the pattern formed by the multiples of 12. Students can write patterns for other students to follow and color. For example, 2r + 3d.

Multiples of 9

Activity and Game Notes and Answers

Further ideas for looking at patterns within multiple patterns:

- A useful extension to this idea is to provide students with a blank 10x10 grid and ask them to shade in squares according to a rule. The shaded grid may then be passed on to a friend to work out the pattern. For example, a pattern that starts in the top left corner and moves across two and down one (as a knight moves in chess), will produce the same pattern as the multiples of twelve.

Multiple Patterns
Page 61

Answers

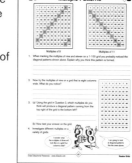

1. The multiples of nine form a diagonal pattern running from the top right to the bottom left of the grid because the grid is ten columns wide. The multiples of nine are related to the multiples of ten in the following way;

   $1 \times 9 = 1 \times 10 - 1$

   $2 \times 9 = 2 \times 10 - 2$

   $3 \times 9 = 3 \times 10 - 3$

   $4 \times 9 = 4 \times 10 - 4$

   $5 \times 9 = 5 \times 10 - 5$ and so on;

   hence, the multiples of nine step back one square, two squares, three squares and so on.

   A similar pattern exists with the multiples of eleven;

   $1 \times 11 = 1 \times 10 + 1$

   $2 \times 11 = 1 \times 10 + 2$

   $3 \times 11 = 1 \times 10 + 3$ and so on.

   Therefore, the diagonal pattern moves from the top left to the bottom right of the grid, stepping in an extra square as it moves down the rows.

2. When placed on a grid that is eight columns wide the multiples of nine form a similar diagonal pattern to the multiples of eleven on a grid ten columns wide. This is because:

   $1 \times 9 = 1 \times 8 + 1$

   $1 \times 9 = 1 \times 8 + 2$

   $1 \times 9 = 1 \times 8 + 3$ and so on.

3. The multiples of seven.

More Multiple Patterns
Page 62

Answers

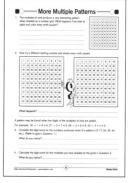

1. A diagonal pattern running from the top right of the grid to the bottom left is created, beginning at eight.

2. More diagonal patterns are formed.

3. The digit sums for 8, 17, 26, 35 and so on all have the sum of eight.

4. The digit sum for the other multiples will depend on the starting number. For example, if you start at five and then add nine, the following sequence is formed; 5, 14, 23, 32, 41, 50 and so on. The digit sum is always five.

Multiple Multiple Patterns – 1
Page 63

Answers

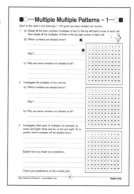

1. (a) Multiples of 2–2, 4, 6, 8, 10, 12, 14, 16, 18, 20, 22, 24, 26, 28, 30, 32, 34, 36, 38, 40, 42, 44, 46, 48, 50, 52, 54, 56, 58, 60, 62, 64, 66, 68, 70, 72, 74, 76, 78, 80, 82, 84, 86, 88, 90, 92, 94, 96, 98, 100

   Multiples of 3–3, 6, 9, 12, 15, 18, 21, 24, 27, 30, 33, 36, 39, 42, 45, 48, 51, 54, 57, 60, 63, 66, 69, 72, 75, 78, 81, 84, 87, 90, 93, 96, 99

   (b) When all the multiples of two and three are shaded, the numbers shaded twice are the multiples of six–6, 12, 18, 24, 30, 36, 42 and so on. The multiples of six are shaded because two and three are factors of six.

   (c) They are odd numbers that are not multiples of three. (All of the even numbers are multiples of two.)

2. Multiples of 4–4, 8, 16, 20, 24, 28, 32, 36, 40, 44, 48, 52, 56, 60, 64, 68, 72, 76, 80, 84, 88, 92, 96

   Multiples of 6–6, 12, 18, 24, 30, 36, 42, 48, 54, 60, 66, 72, 78, 84, 90, 96

   (a) 12, 24, 36, 48, 60 and so on.

   The numbers that are shaded twice are multiples of 12. This is because 12 is the lowest common multiple (LCM) of four and six.

   (b) They are not multiples of four or six.

Activity and Game Notes and Answers

Multiple Multiple Patterns – 2
Page 64

Answers

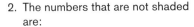
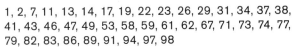

1. Multiples of 3–3, 6, 9, 12, 15, 18, 21, 24, 27, 30, 33, 36, 39, 42, 45, 48, 51, 54, 57, 60, 63, 66, 69, 72, 75, 78, 81, 84, 87, 90, 93, 96, 99

    Multiples of 4–4, 8, 12, 16, 20, 24, 28, 32, 36, 40, 44, 48, 52, 56, 60, 64, 68, 72, 76, 80, 84, 88, 92, 96

    Multiples of 5–5, 10, 15, 20, 25, 30, 35, 40, 45, 50, 55, 60, 65, 70, 75, 80, 85, 90, 95, 100

2. The numbers that are not shaded are:

    1, 2, 7, 11, 13, 14, 17, 19, 22, 23, 26, 29, 31, 34, 37, 38, 41, 43, 46, 47, 49, 53, 58, 59, 61, 62, 67, 71, 73, 74, 77, 79, 82, 83, 86, 89, 91, 94, 97, 98

3. In each row there is a pair of adjacent numbers that are not marked.

4. (a) The number shaded in all three corners is 60.

    (b) This is the only number between 1 and 100 which has factors of three, four and five.

    Note: 3 x 4 x 5 = 60. Sixty is the lowest common multiple (LCM) of three, four and five.

5. 120 and 180 would be shaded in three corners.

6. Multiples of 2–2, 4, 6, 8, 10, 12, 14, 16, 18, 20, 22, 24, 26, 28, 30, 32, 34, 36, 38, 40, 42, 44, 46, 48, 50, 52, 54, 56, 58, 60, 62, 64, 66, 68, 70, 72, 74, 76, 78, 80, 82, 84, 86, 88, 90, 92, 94, 96, 98, 100

    Multiples of 4–4, 8, 12, 16, 20, 24, 28, 32, 36, 40, 44, 48, 52, 56, 60, 64, 68, 72, 76, 80, 84, 88, 92, 96

    Multiples of 6–6, 12, 18, 24, 30, 36, 42, 48, 54, 60, 66, 72, 78, 84, 90, 96

    Multiples of 8–8, 16, 24, 32, 40, 48, 56, 64, 72, 80, 88, 96

7. 4, 6, 18, 20, 28, 30, 42, 44, 52, 54, 66, 68, 76, 78, 90, 92 and 100 have two squares shaded.

    Note: These numbers come in even-numbered pairs. It is likely that 102 will have two squares shaded.

8. 8, 12, 16, 32, 36, 40, 56, 60, 64, 80, 84, 88 have three squares shaded.

    Note: These numbers come in triples with each number four more than the next.

9. 24, 48, 72, 96 have four squares shaded.

10. These numbers are multiples of 24, which is the lowest common multiple (LCM) of 2, 4, 6 and 8. The students could predict and test whether 120 is the next number by using a 101-200 grid.

Cracking the Code – 1 & 2
Pages 65 – 66

Answers

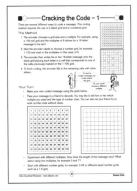

1. Multiples of 5 – I cannot wait for recess.

2. Multiples of 7 – Help me with this.

3. Multiples of 4 and 6 – Thousand.

Digit Sums
Page 67

Answers

1.

1	2	3	4	5	6	7	8	9	1
2	3	4	5	6	7	8	9	1	2
3	4	5	6	7	8	9	1	2	3
4	5	6	7	8	9	1	2	3	4
5	6	7	8	9	1	2	3	4	5
6	7	8	9	1	2	3	4	5	6
7	8	9	1	2	3	4	5	6	7
8	9	1	2	3	4	5	6	7	8
9	1	2	3	4	5	6	7	8	9
1	2	3	4	5	6	7	8	9	1

2. Teacher check

Combining Columns
Page 68

Answers

1. fifth column

2. The answer always appears in the fifth column.

3. seventh column

4. ninth column

5-6. The answer when adding column two and four will be found in column six, three and seven in ten, six and seven in column three (i.e. 13-10) and so on.

7. The answer to doubling a number in column four may be found in column eight.

8. Similar patterns may be found as when doubling.

Shape Shifter
Page 69

Answers

1. 24

2. 27

3. 30, 33, 36, 39, 42, 45, 48

4. The totals increase by three each time.

5. 24, 54, 84, 114, 144, 174, 204, 234, 264.

6. The totals increase by 30 each time.

7. 24, 57, 90, 123. The total increases by 33 as the "L" shape moves in a diagonal across the page.

Windows and Overlays – 1
Page 70

Answers

1. The two numbers are 10 apart.

2. The two numbers are one apart.

3. Window (B) produces two numbers eight apart.

   Window (C) produces two numbers 17 apart.

   Window (D) produces two numbers nine apart.

   Window (E) produces two numbers 18 apart.

4. Window (B) when rotated 90° produces two numbers that are 21 apart.

   Window (C) when rotated 90° produces two numbers that are 32 apart.

   Window (D) when rotated 90° produces two numbers that are 11 apart.

   Window (E) when rotated 90° produces two numbers that are 22 apart.

Note: A dot appears in the top left box of each window for correct placement.

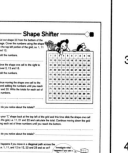

Windows and Overlays – 2
Page 71

Answers

1. + 11 window

Note: The position on the window does not matter but the relative position of the cut out does.

2. + 3 window

3. + 8 window

4.+ 12 window

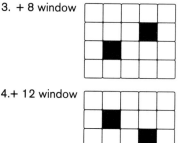

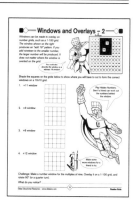

Challenge: When the multiples of 9 are cut out and the grid overlaid and rotated 90° a +11 window is produced.

Note: Some students may find it difficult to cut out the multiples of nine so you may wish to prepare an overlay for the overhead projector.

Odd Squares
Page 72

Answers

Multiplying the center number by four will produce the same total as the sum of the four corner numbers. The same relationship exists in a 5x5 square and a 7x7 square.

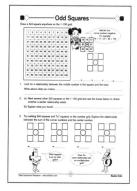

Reversals
Page 73

Answers

The result answer will always be a multiple of nine.

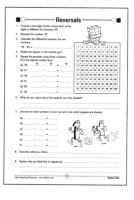

Revealing Rectangles
Page 74

Answers

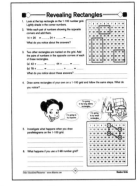

2. 14 + 26 = 40, 24 + 16 = 40
   Both answers are the same.

3. (a) 42 + 94 = 136,
           44 + 92 = 136

   (b) 78 + 100 = 178,
           80 + 98 = 178

In each case, the pairs of numbers in opposite corners of the rectangle add to the same amount.

4. In each case, pairs of numbers in the opposite corners of a rectangle will add to the same amount.

5. Pairs of numbers in the opposite corners of a parallelogram will also add to the same amount.

6. The same thing will occur if you use a 0-99 grid.

Four Squares
Page 75

Answers

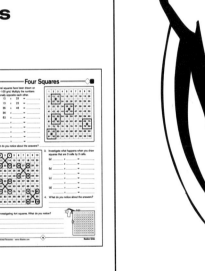

1. (a) 12 x 23 = 276,
           13 x 22 = 286

   (b) 35 x 46 = 1,610,
           36 x 45 = 1,620

   (c) 52 x 63 = 3,276,
           53 x 62 = 3,286

   (d) 69 x 80 = 5,520,
           70 x 79 = 5,530

   (e) 81 x 92 = 7,452,
           82 x 91 = 7,462

2. In each case, the answers differ by ten.

3. (a) 2 x 24 = 48, 4 x 22 = 88
   (b) 26 x 48 = 1,248, 28 x 46 = 1,288
   (c) 52 x 74 = 3,848, 54 x 72 = 3,888
   (d) 78 x 100 = 7,800, 80 x 98 = 7,840

4. In each case, the answers differ by 40.

5. In each case, the answers differ by 90.

Didax Educational Resources ~ www.didaxinc.com

Number Grids

Number Grid Games and Activities

pages 41 - 75

In this section you will find a collection of games and activities for students to use with number grids.

Some games and activities need to be photocopied for each student. Others can be photocopied once, glued on card stock and laminated, or placed in a plastic sleeve for student use.

The majority of the games and activities use the 1-100 number grid. However, in most cases, any grid configuration can be substituted. Refer to Section 1 for templates of completed or blank grids.

Other equipment needed by the students, such as dice, cards, and cubes, is listed on the appropriate pages.

The games and activities are arranged in order of difficulty.

S
e
c
t
i
o
n

3

Cut and Paste

You will need:

- scissors
- glue

Follow the directions below to make a number track and number roll.

1	2	3	4	5	6	7	8	9	10	●
11	12	13	14	15	16	17	18	19	20	●
21	22	23	24	25	26	27	28	29	30	●
31	32	33	34	35	36	37	38	39	40	●
41	42	43	44	45	46	47	48	49	50	●
51	52	53	54	55	56	57	58	59	60	●
61	62	63	64	65	66	67	68	69	70	●
71	72	73	74	75	76	77	78	79	80	●
81	82	83	84	85	86	87	88	89	90	●
91	92	93	94	95	96	97	98	99	100	

Number Track

- Cut out the number grid around each row.
- Glue the rows together in order at each dot.
- Spread the track out and see how far it extends.
- Play a racing track game with a friend. All you need is a die and some counters.
- Each player takes turns rolling the die and moving along the track the number of spaces indicated by the die. The first player to reach the end is the winner.

Number Roll

- Cut out the number grid around each column.
- Glue the columns together in order at each dot.
- When you have finished gluing the columns, wind your strip around the end of your pencil to make a number roll.

1	11	21	31	41	51	61	71	81	91
2	12	22	32	42	52	62	72	82	92
3	13	23	33	43	53	63	73	83	93
4	14	24	34	44	54	64	74	84	94
5	15	25	35	45	55	65	75	85	95
6	16	26	36	46	56	66	76	86	96
7	17	27	37	47	57	67	77	87	97
8	18	28	38	48	58	68	78	88	98
9	19	29	39	49	59	69	79	89	99
10	20	30	40	50	60	70	80	90	100
●	●	●	●	●	●	●	●	●	●

Count 'n Color

You will need:

- a calculator
- a 1-100 grid (enlarged)
- cubes or tiles to cover numbers on the grid

1	2	3	4	5	6	7	8	9	10
11	12	13	14	15	16	17	18	19	20
21	22	23	24	25	26	27	28	29	30
31	32	33	34	35	36	37	38	39	40
41	42	43	44	45	46	47	48	49	50
51	52	53	54	55	56	57	58	59	60
61	62	63	64	65	66	67	68	69	70
71	72	73	74	75	76	77	78	79	80
81	82	83	84	85	86	87	88	89	90
91	92	93	94	95	96	97	98	99	100

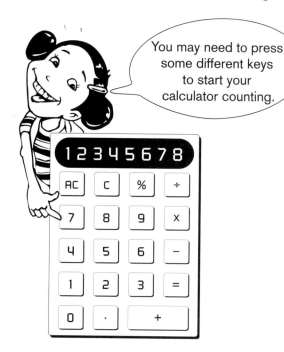

You may need to press some different keys to start your calculator counting.

1. Enter $+ 5 = = =$ into your calculator. What do you notice?

_____

_____

2. Clear your calculator and then enter $+ 4 = = =$ into your calculator. Place a tile on your number grid each time another four is added. What do you notice?

_____

_____

- Continue until you reach the end of the grid.
- Leave the tiles on the grid.

3. Clear your calculator and enter $+ 8 = = =$. Leave tiles on the grid to show the numbers that appear each time another eight is added. Remove tiles from the numbers that do not appear on the display. What do you notice?

_____

_____

4. Remove all of the tiles from the grid and clear your calculator. Press $+ 6 = = =$ and place tiles on the grid to show which numbers appear. Continue until you reach the end of the grid, then clear the calculator and enter $+ 3 = = =$. Place a tile on the grid each time another three is added. Write about what you discover.

_____

_____

_____

Didax Educational Resources ~ www.didaxinc.com

Hidden Numbers

You will need:

- a partner
- a 1-100 grid (enlarged)
- cubes or tiles to cover numbers on the grid

How to Play

1. Hide a Number

One player covers a number on the grid. The other player has to guess the number and explain how he or she worked it out. For example, if the number was 44, you could reason that it is one less than 45 or ten more than 34 etc. Play for several rounds, taking turns.

1	2	3	4	5	6	7	8	9	10
11	12	13	14	15	16	17	18	19	20
21	22	23	24	25	26	27	28	29	30
31	32	33	34	35	36	37	38	39	40
41	42	43	■	45	46	47	48	49	50
51	52	53	54	55	56	57	58	59	60
61	62	63	64	65	66	67	68	69	70
71	72	73	74	75	76	77	78	79	80
81	82	83	84	85	86	87	88	89	90
91	92	93	94	95	96	97	98	99	100

2. Letters and Shapes

One player links tiles to make a letter shape, such as "L" or "H," or a shape such as a square or rectangle, on the grid. The other player has to identify all the numbers covered by this shape. The player needs to explain how he or she worked out which numbers were covered. Play for several rounds, taking turns.

1	2	3	4	5	6	7	8	9	10
11	12	13	14	15	16	17	18	19	20
21	22	23	24	25	26	27	28	29	30
31	32	33	■	35	36	37	38	39	40
41	42	43	■	45	46	47	48	49	50
51	52	53	■	55	56	57	58	59	60
61	62	63	■	65	66	67	68	69	70
71	72	73	■	■	■	77	78	79	80
81	82	83	84	85	86	87	88	89	90
91	92	93	94	95	96	97	98	99	100

3. Find the Number

Cover all the numbers on the grid. One player chooses a number and tells it to the other player. That player lifts a tile to reveal the chosen number. If the first tile lifted doesn't reveal the chosen number, leave that tile off and keep lifting tiles until the chosen number is revealed.

Write how many tiles were lifted and replace all the tiles. Play for several rounds, taking turns. The player who lifts the least number of tiles over several rounds is the winner.

Missing Numbers

1. The following 1-100 number grid has been partially completed. Insert the following numbers onto the grid, leaving all the other cells blank.

7, 98, 42, 13, 56, 25, 34, 70, 81, 79

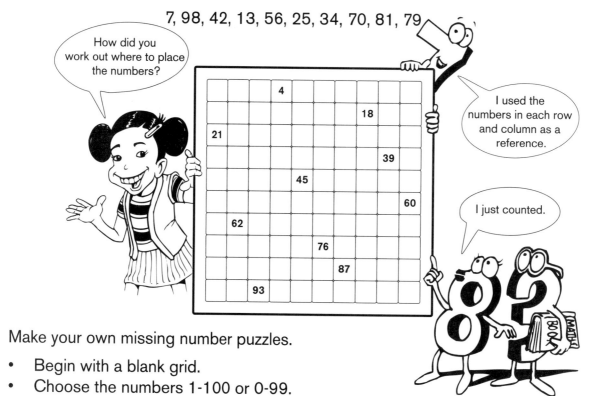

2. Make your own missing number puzzles.

- Begin with a blank grid.
- Choose the numbers 1-100 or 0-99.
- Now enter ten clues. There should be one clue (number) in each column and row of the grid.
- Now choose ten numbers to be placed onto the grid and write them beside the grid.
- Cut out and give your grid to a friend to complete.
- Repeat the directions using the second grid and cut it out and give it to a different friend to complete.

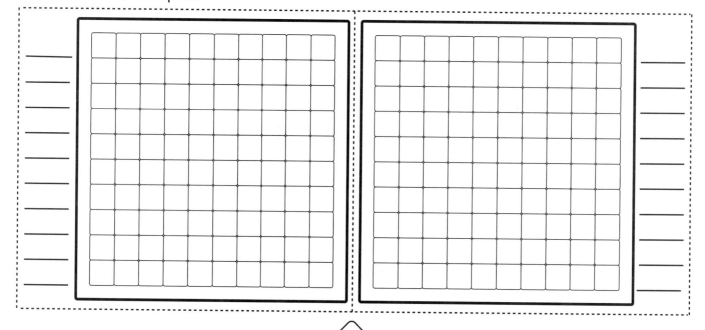

Number Grid Games

These games need two players. So find yourself a partner!

You will need:

- a 1-100 grid
- two 10-sided dice numbered 0-9 *or* two sets of 10 cards numbered 0-9 *or* two spinners with sections numbered 0-9
- cubes or tiles in two colors to cover the numbers on the grid (each player should have cubes or tiles of one color)

How to Play

Decide with your partner whether you want to use dice, cards, or spinners to play one of the following games. Once you have played for awhile, play one of the other games.

Roll and Place

1. The player who rolls the highest number goes first.
2. The first player rolls both dice and uses the digits shown to produce a number.
3. If, for example, a two and a five are thrown, then either the number 25 or 52 can be chosen. The player covers one of these numbers on the grid.
4. The other player takes a turn.
5. The aim of the game is to be the first player to cover three numbers in a row; horizontally, vertically, or diagonally.

Pick and Place

1. Place one deck of cards face down. The player who chooses the card with the highest number goes first.
2. Each card deck is shuffled separately and placed face down. The first player turns over the top card from each deck and uses the digits shown to produce a number.
3. If, for example, a seven and a nine are picked, then number 79 or 97 can be chosen. The player covers one of these numbers on the grid.
4. The other player takes a turn.
5. The aim of the game is to be the first player to cover three numbers in a row: horizontally, vertically, or diagonally.

Spin and Place

1. Each player spins a spinner. The player who spins the highest number goes first.
2. The first player spins both spinners and uses the digits shown to produce a number.
3. If, for example, a three and a zero are thrown, then either the number 30 or 3 can be chosen. The player covers one of these numbers on the grid.
4. The other player takes a turn.
5. The aim of the game is to be the first player to cover three numbers in a row: horizontally, vertically, or diagonally.

Jigsaws – 1

Number grid jigsaws are easy to make. All you need to do is choose a grid and cut it into pieces. The pieces can then be reassembled.

1. Cut out the pieces below and reform them into a 1-100 number grid.

2. Using the 1-100 grid on the right, make your own jigsaw for a friend to solve.

1	2	3	4	5	6	7	8	9	10
11	12	13	14	15	16	17	18	19	20
21	22	23	24	25	26	27	28	29	30
31	32	33	34	35	36	37	38	39	40
41	42	43	44	45	46	47	48	49	50
51	52	53	54	55	56	57	58	59	60
61	62	63	64	65	66	67	68	69	70
71	72	73	74	75	76	77	78	79	80
81	82	83	84	85	86	87	88	89	90
91	92	93	94	95	96	97	98	99	100

If you want to make it easier you can form your jigsaw on top of a number grid.

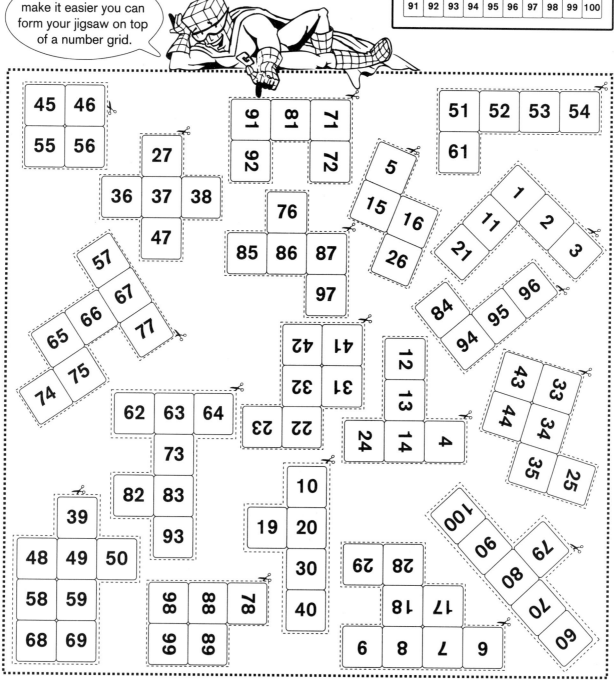

More Missing Numbers

1. Fill in the missing numbers in each of the pieces of the 1-100 number grid.

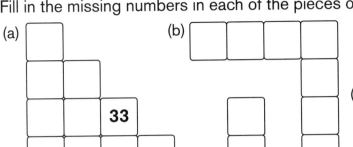

(a) (b) (c) (d)

33 68 45 73

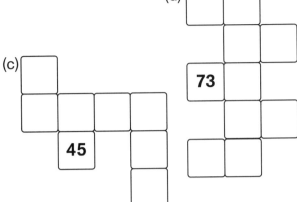

(e)

56

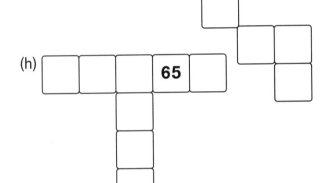

(h) 65

(f) 47 (g) 27

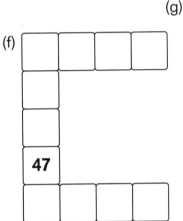

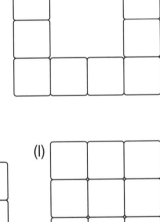

(i) 69 (j) (k) 88 (l) 41

62

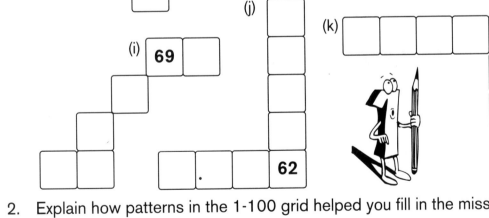

2. Explain how patterns in the 1-100 grid helped you fill in the missing numbers.

_____

_____

_____

3. Make up a missing number puzzle for a friend to solve. Begin by cutting up a grid and then whiting out some of the numbers. Cut out pieces in the shape of letters.

Didax Educational Resources ~ www.didaxinc.com

Number Grids

You can make a more difficult jigsaw by blocking out some of the numbers on each piece. The jigsaw below was made with a 1-100 grid. Sections were cut out and some numbers left blank.

1. Cut out the pieces below and reform them into the grid.

2. Using the 1-100 grid on the right, make your own jigsaw by cutting pieces and whiting some numbers out.

Challenge: Use another grid, such as the spiral grid.

1	2	3	4	5	6	7	8	9	10
11	12	13	14	15	16	17	18	19	20
21	22	23	24	25	26	27	28	29	30
31	32	33	34	35	36	37	38	39	40
41	42	43	44	45	46	47	48	49	50
51	52	53	54	55	56	57	58	59	60
61	62	63	64	65	66	67	68	69	70
71	72	73	74	75	76	77	78	79	80
81	82	83	84	85	86	87	88	89	90
91	92	93	94	95	96	97	98	99	100

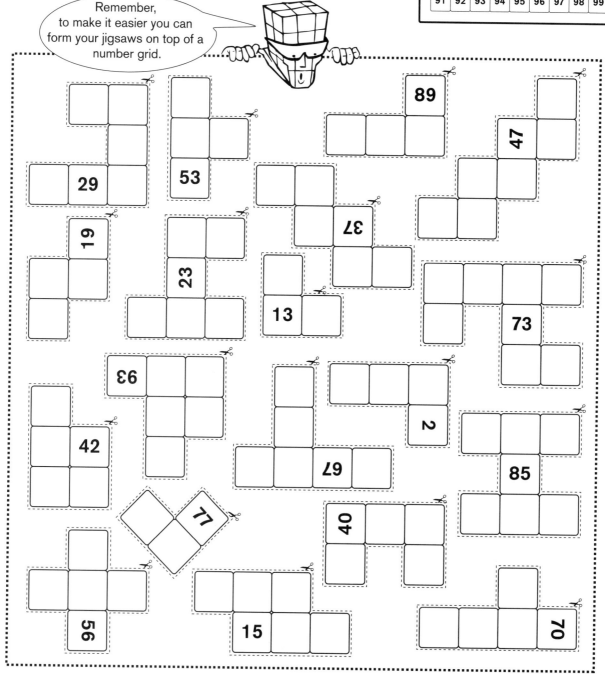

Placing for Points

This game is for 2-4 players.

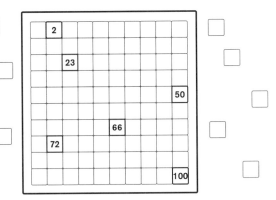

You will need
- set of numbers 1-100 (to fit grid cells)
- pencil and paper to tally scores
- a completed 1-100 grid

How to Play
1. Place the numbered pieces face down.
2. Each player takes a turn, picking up a piece and placing it on the grid.
3. Players score a point for each piece that is correctly placed. Look at the completed grid if a player is thought to be incorrect. (The completed grid should be face down and only looked at when necessary.)
4. The game continues for a set length of time or until all the pieces are placed correctly.

Number Grids

Number Spirals

There are many different spiral number grids that can be made. The number spiral on the right begins in the center and spirals in a clockwise direction until it reaches the outside of the grid.

1. Make a check mark in the cell where you think the number 100 will be entered. Explain why you chose this spot.

    _____

    _____

2. Make a cross in the spot where you think the number 50 will be. Explain why you chose this spot.

    _____

    _____

3. What numbers do you think will end up in the shaded cell and the black cell?

    Shaded cell: _____   Black cell: _____

    Explain. _____

4. Complete the spiral to check your thinking.

    -

    The number spiral to the left begins in the top left corner and spirals in an counterclockwise direction.

5. Make a check where you think the number 100 will be entered. Explain why you chose this spot.

    _____

    _____

6. Make a cross where you think the number 50 will be. Explain why you chose this spot.

    _____

    _____

7. What numbers do you think will end up in the shaded cell and the black cell?

    Shaded cell: _____   Black cell: _____

    Explain. _____

8. Make up your own number spiral using a different type of spiral grid.

Didax Educational Resources ~ www.didaxinc.com

All the Right Moves – 1

1. Follow these routes around the 1-100 grid and write the finishing number. For example, 15 → ↓ →→ ↓ follows the route shown on the grid and finishes at 38.

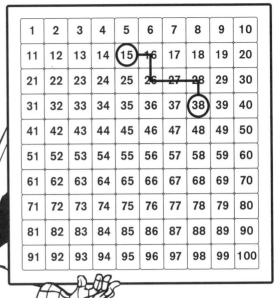

1	2	3	4	5	6	7	8	9	10
11	12	13	14	15	16	17	18	19	20
21	22	23	24	25	26	27	28	29	30
31	32	33	34	35	36	37	38	39	40
41	42	43	44	45	46	47	48	49	50
51	52	53	54	55	56	57	58	59	60
61	62	63	64	65	66	67	68	69	70
71	72	73	74	75	76	77	78	79	80
81	82	83	84	85	86	87	88	89	90
91	92	93	94	95	96	97	98	99	100

   Try these.

   (a) 54   →→ ↑ →   _____

   (b) 100   ↑ ↑ ←← ↑   _____

   (c) 59   ←← ↑ ↑   _____

   (d) 72   → ↑ ↑ →→   _____

   (e) 44   →→ ↑ ← ↓ ←   _____

   What do you notice about question 1(e)?

   There must be a quicker way of writing ↑ ↑ ↑

   (f) 13 ↓ ↓ ↓ →→→   _____

   (g) 62 ← ↑ ↑ ↑   _____

   (h) 86 ↑ ↑ ↑ ←←←   _____

   (i) 35 ↓ ↓ ↓ →→   _____

   (j) 67 →→ ↓ ↓ ↓   _____

2. Design some routes that will take you from one number to another.

   (a) 15 _____ _____ _____ 33

   (b) 76 _____ _____ _____ _____ 48

   (c) 22 _____ _____ _____ _____ _____ 55

   (d) 61 _____ _____ _____ 55

   (e) 55 _____ _____ _____ _____ _____ 22

   What do you notice about question 2 parts (c) and (e)?

Make some routes for a friend to try.

You might like to add some more arrow symbols.

3. What do you think each of these arrow symbols might mean?

(a) _____ (b) _____ (c) _____ (d) _____ (e) _____ (f) _____

Number Grids

All the Right Moves – 2

So far you have used arrows to describe moves on a standard 1-100 grid numbered from the top down. For example:

- The arrow → means move one place to the right or add one.

- The arrow ↑ means move directly up one square or subtract 10.

Do the arrows mean the same thing on different number grids? Investigate for other number grids based on multiples of 2, 3, 4, etc. For example, on a 2-200 number grid → still means move one cell or square to the right, but now it means you will be adding two instead of one.

2	4	6	8	10	12	14	16	18	20
22	24	26	28	30	32	34	36	38	40
42	44	46	48	50	52	54	56	58	60
62	64	66	68	70	72	74	76	78	80
82	84	86	88	90	92	94	96	98	100
102	104	106	108	110	112	114	116	118	120
122	124	126	128	130	132	134	136	138	140
142	144	146	148	150	152	154	156	158	160
162	164	166	168	170	172	174	176	178	180
182	184	186	188	190	192	194	196	198	200

1. What will ↑ mean on a 2-200 number grid?

   _____

2. Describe what each of these arrows will mean on a 2-200 number grid.

   (a) ← _____   (d) ↓ _____

   (b) ↙ _____   (e) ↘ _____

   (c) ↖ _____   (f) ↗ _____

3. What would these arrows mean on a 5-500 grid?

   (a) ← _____   (d) ↓ _____

   (b) ↙ _____   (e) ↘ _____

   (c) ↖ _____   (f) ↗ _____

5	10	15	20	25	30	35	40	45	50
55	60	65	70	75	80	85	90	95	100
105	110	115	120	125	130	135	140	145	150
155	160	165	170	175	180	185	190	195	200
205	210	215	220	225	230	235	240	245	250
255	260	265	270	275	280	285	290	295	300
305	310	315	320	325	330	335	340	345	350
355	360	365	370	375	380	385	390	395	400
405	410	415	420	425	430	435	440	445	450
455	460	465	470	475	480	485	490	495	500

4. Choose another 10x10 number grid, name it and describe what the following arrows mean.

   Grid: _____

   (a) → _____   (e) ← _____   (i) ↓ _____

   (b) ↑ _____   (f) ↙ _____   (j) ↘ _____

   (c) ↗ _____   (g) ↖ _____   (k) ↷ _____

   (d) ⇓ _____   (h) ↗ _____   (l) ⇉ _____

5. Describe what you notice when the grid numbering system changes. _____
   _____
   _____

All the Right Moves – 3

So far you have used arrows to describe moves on various 100-cell number grids made up of ten columns and ten rows. Investigate what happens if you use arrows to describe moves on different sized grids. For example, on a 10x10, 1-100 number grid an ↓ arrow means move down one cell or add ten.

1. What would these arrows mean on a 5 x 5 grid numbered 1-25?

    (a) ↓ _____

    (b) ↗ _____

    (c) ↙ _____

There are many different-sized grids to investigate.

1	2	3	4	5
6	7	8	9	10
11	12	13	14	15
16	17	18	19	20
21	22	23	24	25

2. Choose a grid to investigate. Describe the grid you are investigating and what each of the following arrows mean when applied to the grid.

    Grid: _____

(a) _____

(b) _____

(c) _____

(d) _____

(e)  _____

(f) _____

(g)  _____

(h)  _____

(i)  _____

(j)  _____

(k) _____

(l) _____

(m)  _____

(n)  _____

(o) _____

3. Describe what you notice when the grid size changes.

    _____

    _____

    _____

    _____

    _____

Grid Race

This is a game for 2-4 players.

You will need

- the same grid for each player
- a set of arrow cards
- a marker for each player (colored counter, cube, or tile)
- a pencil or highlighter

How to Play

Grid Race

1. Place the arrow cards face down in a pile.

2. Each player begins with a marker in the top left corner of the grid.

3. In turn, each player draws a card from the pile and moves his or her marker. (If a player can not move, simply pick up another card.)

4. After ten turns, the player whose marker is on the highest number is the winner.

Each player should record the route traveled on his or her grid.

Variations

- Use a six-sided die marked with +11, −1, +9, +10, +11, and −10. (Other markings can be used.)
- Use a spinner instead of a die.
- Use a blank grid and write the numbers on the grid as you move along the route.
- Start at the top right of the grid and only use arrow cards that subtract.
- Begin in the middle of the grid.
- Use another grid, such as a Down and Up grid.

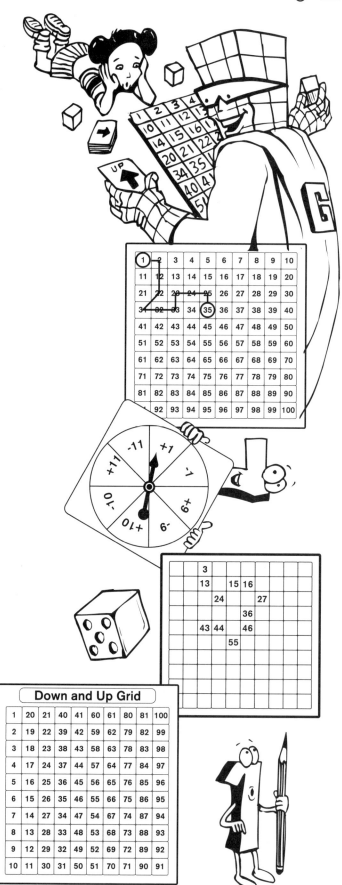

Didax Educational Resources ~ www.didaxinc.com

Think of a Number

You can play this game in a small group (3-5 players is best).

You will need:

- the same grid for each player (1-100, 0-99, etc.)
- a pencil or highlighter

How to Play

1. One player thinks of a number on the grid and marks it on his or her copy.

2. The other players take turns asking questions to find the number.

   For example,

3. Each player shades in the grid according to the answer given. This keeps track of what the answer cannot be and which questions are better to ask than others.

4. When the answer is found, another player chooses a number on a new grid and the game starts over.

5. You should be able to find any number on the grid by asking seven questions or less!

Number Grids

Eratosthenes' Sieve

1. Starting with the 1-100 grid,

   - Cross out the number 1.

   - Circle the number 2 and then cross out all the multiples of 2.

   - Circle 3 and then cross out all the multiples of 3.

   - Circle 5 and then cross out all the multiples of 5.

   - Circle 7 and then cross out all the multiples of 7.

The set of numbers left are called prime numbers.

1	2	3	4	5	6	7	8	9	10
11	12	13	14	15	16	17	18	19	20
21	22	23	24	25	26	27	28	29	30
31	32	33	34	35	36	37	38	39	40
41	42	43	44	45	46	47	48	49	50
51	52	53	54	55	56	57	58	59	60
61	62	63	64	65	66	67	68	69	70
71	72	73	74	75	76	77	78	79	80
81	82	83	84	85	86	87	88	89	90
91	92	93	94	95	96	97	98	99	100

I had 2, 3, 5, 7, 11, 13, 17, 19, 23, 29, 31, 37, 41, 43, 47, 53, 59, 61, 67, 71, 73, 79, 83, 89 and 97 left.

2. After the first row, in which columns can prime numbers be found?

_____

_____

3. Now try the same thing on a number grid that is six cells wide. What do you notice?

_____

_____

_____

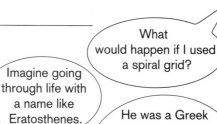

What would happen if I used a spiral grid?

Imagine going through life with a name like Eratosthenes.

Who was Eratosthenes?

He was a Greek mathematician who is credited with being the first to devise a special method. It finds all of the prime numbers less than 100.

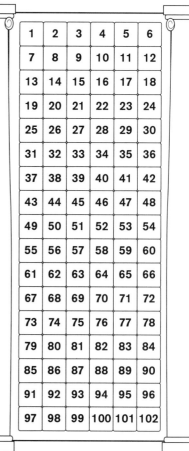

1	2	3	4	5	6
7	8	9	10	11	12
13	14	15	16	17	18
19	20	21	22	23	24
25	26	27	28	29	30
31	32	33	34	35	36
37	38	39	40	41	42
43	44	45	46	47	48
49	50	51	52	53	54
55	56	57	58	59	60
61	62	63	64	65	66
67	68	69	70	71	72
73	74	75	76	77	78
79	80	81	82	83	84
85	86	87	88	89	90
91	92	93	94	95	96
97	98	99	100	101	102

Number Grids

Didax Educational Resources ~ www.didaxinc.com

Mark the multiples of 2, 3, 4, 5, 6 and 7 on the 1-100 grids provided.

What is a multiple?

The multiples of two would be 2, 4, 6, 8, etc.

1	2	3	4	5	6	7	8	9	10
11	12	13	14	15	16	17	18	19	20
21	22	23	24	25	26	27	28	29	30
31	32	33	34	35	36	37	38	39	40
41	42	43	44	45	46	47	48	49	50
51	52	53	54	55	56	57	58	59	60
61	62	63	64	65	66	67	68	69	70
71	72	73	74	75	76	77	78	79	80
81	82	83	84	85	86	87	88	89	90
91	92	93	94	95	96	97	98	99	100

Multiples of 2

1	2	3	4	5	6	7	8	9	10
11	12	13	14	15	16	17	18	19	20
21	22	23	24	25	26	27	28	29	30
31	32	33	34	35	36	37	38	39	40
41	42	43	44	45	46	47	48	49	50
51	52	53	54	55	56	57	58	59	60
61	62	63	64	65	66	67	68	69	70
71	72	73	74	75	76	77	78	79	80
81	82	83	84	85	86	87	88	89	90
91	92	93	94	95	96	97	98	99	100

Multiples of 3

1	2	3	4	5	6	7	8	9	10
11	12	13	14	15	16	17	18	19	20
21	22	23	24	25	26	27	28	29	30
31	32	33	34	35	36	37	38	39	40
41	42	43	44	45	46	47	48	49	50
51	52	53	54	55	56	57	58	59	60
61	62	63	64	65	66	67	68	69	70
71	72	73	74	75	76	77	78	79	80
81	82	83	84	85	86	87	88	89	90
91	92	93	94	95	96	97	98	99	100

Multiples of 4

1	2	3	4	5	6	7	8	9	10
11	12	13	14	15	16	17	18	19	20
21	22	23	24	25	26	27	28	29	30
31	32	33	34	35	36	37	38	39	40
41	42	43	44	45	46	47	48	49	50
51	52	53	54	55	56	57	58	59	60
61	62	63	64	65	66	67	68	69	70
71	72	73	74	75	76	77	78	79	80
81	82	83	84	85	86	87	88	89	90
91	92	93	94	95	96	97	98	99	100

Multiples of 5

1	2	3	4	5	6	7	8	9	10
11	12	13	14	15	16	17	18	19	20
21	22	23	24	25	26	27	28	29	30
31	32	33	34	35	36	37	38	39	40
41	42	43	44	45	46	47	48	49	50
51	52	53	54	55	56	57	58	59	60
61	62	63	64	65	66	67	68	69	70
71	72	73	74	75	76	77	78	79	80
81	82	83	84	85	86	87	88	89	90
91	92	93	94	95	96	97	98	99	100

Multiples of 6

1	2	3	4	5	6	7	8	9	10
11	12	13	14	15	16	17	18	19	20
21	22	23	24	25	26	27	28	29	30
31	32	33	34	35	36	37	38	39	40
41	42	43	44	45	46	47	48	49	50
51	52	53	54	55	56	57	58	59	60
61	62	63	64	65	66	67	68	69	70
71	72	73	74	75	76	77	78	79	80
81	82	83	84	85	86	87	88	89	90
91	92	93	94	95	96	97	98	99	100

Multiples of 7

What happens if you use a different number grid?

Try to describe each pattern.

Marking Multiples – 2

Mark the multiples of 8, 9, 10, 11 and 12 on the 1-100 grids provided.

What is a multiple?

The multiples of eight would be 8, 16, 24, etc.

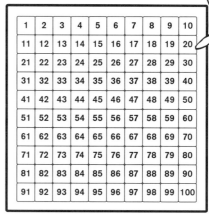

1	2	3	4	5	6	7	8	9	10
11	12	13	14	15	16	17	18	19	20
21	22	23	24	25	26	27	28	29	30
31	32	33	34	35	36	37	38	39	40
41	42	43	44	45	46	47	48	49	50
51	52	53	54	55	56	57	58	59	60
61	62	63	64	65	66	67	68	69	70
71	72	73	74	75	76	77	78	79	80
81	82	83	84	85	86	87	88	89	90
91	92	93	94	95	96	97	98	99	100

Multiples of 8

1	2	3	4	5	6	7	8	9	10
11	12	13	14	15	16	17	18	19	20
21	22	23	24	25	26	27	28	29	30
31	32	33	34	35	36	37	38	39	40
41	42	43	44	45	46	47	48	49	50
51	52	53	54	55	56	57	58	59	60
61	62	63	64	65	66	67	68	69	70
71	72	73	74	75	76	77	78	79	80
81	82	83	84	85	86	87	88	89	90
91	92	93	94	95	96	97	98	99	100

Multiples of 9

1	2	3	4	5	6	7	8	9	10
11	12	13	14	15	16	17	18	19	20
21	22	23	24	25	26	27	28	29	30
31	32	33	34	35	36	37	38	39	40
41	42	43	44	45	46	47	48	49	50
51	52	53	54	55	56	57	58	59	60
61	62	63	64	65	66	67	68	69	70
71	72	73	74	75	76	77	78	79	80
81	82	83	84	85	86	87	88	89	90
91	92	93	94	95	96	97	98	99	100

Multiples of 10

1	2	3	4	5	6	7	8	9	10
11	12	13	14	15	16	17	18	19	20
21	22	23	24	25	26	27	28	29	30
31	32	33	34	35	36	37	38	39	40
41	42	43	44	45	46	47	48	49	50
51	52	53	54	55	56	57	58	59	60
61	62	63	64	65	66	67	68	69	70
71	72	73	74	75	76	77	78	79	80
81	82	83	84	85	86	87	88	89	90
91	92	93	94	95	96	97	98	99	100

Multiples of 11

What happens if you use a different number grid?

1	2	3	4	5	6	7	8	9	10
11	12	13	14	15	16	17	18	19	20
21	22	23	24	25	26	27	28	29	30
31	32	33	34	35	36	37	38	39	40
41	42	43	44	45	46	47	48	49	50
51	52	53	54	55	56	57	58	59	60
61	62	63	64	65	66	67	68	69	70
71	72	73	74	75	76	77	78	79	80
81	82	83	84	85	86	87	88	89	90
91	92	93	94	95	96	97	98	99	100

Multiples of 12

Try to describe each pattern.

Didax Educational Resources ~ www.didaxinc.com

Multiple Patterns

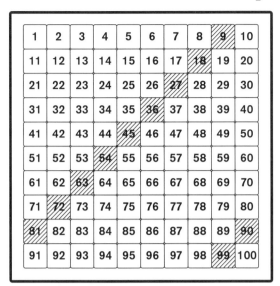

Multiples of 9

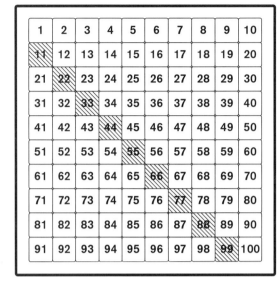

Multiples of 11

1. When marking the multiples of nine and eleven on a 1-100 grid you probably noticed the diagonal patterns shown above. Explain why you think this pattern is formed.

_____

_____

_____

2. Now try the multiples of nine on a grid that is eight columns wide. What do you notice?

_____

_____

_____

1	2	3	4	5	6	7	8
9	10	11	12	13	14	15	16
17	18	19	20	21	22	23	24
25	26	27	28	29	30	31	32
33	34	35	36	37	38	39	40
41	42	43	44	45	46	47	48
49	50	51	52	53	54	55	56
57	58	59	60	61	62	63	64
65	66	67	68	69	70	71	72
73	74	75	76	77	78	79	80
81	82	83	84	85	86	87	88
89	90	91	92	93	94	95	96

3. (a) Using the grid in Question 2, which multiples do you think will produce a diagonal pattern running from the top right of the grid to the bottom left?

_____

_____

(b) Now test your answer on the grid.

4. Investigate different multiples on a variety of grids.

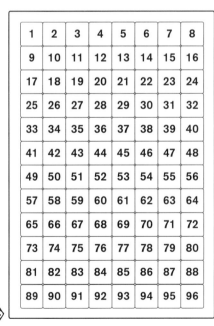

I wonder what the multiples of three will look like on a grid four columns wide?

I am going to look at diagonal patterns like the ones above.

Didax Educational Resources ~ www.didaxinc.com

Number Grids

More Multiple Patterns

1. The multiples of nine produce a very interesting pattern when shaded on a number grid. What happens if we start at eight and color every ninth square?

1	2	3	4	5	6	7	8	9	10
11	12	13	14	15	16	17	18	19	20
21	22	23	24	25	26	27	28	29	30
31	32	33	34	35	36	37	38	39	40
41	42	43	44	45	46	47	48	49	50
51	52	53	54	55	56	57	58	59	60
61	62	63	64	65	66	67	68	69	70
71	72	73	74	75	76	77	78	79	80
81	82	83	84	85	86	87	88	89	90
91	92	93	94	95	96	97	98	99	100

_____

_____

_____

_____

_____

2. Now try a different starting number and shade every ninth square.

1	2	3	4	5	6	7	8	9	10
11	12	13	14	15	16	17	18	19	20
21	22	23	24	25	26	27	28	29	30
31	32	33	34	35	36	37	38	39	40
41	42	43	44	45	46	47	48	49	50
51	52	53	54	55	56	57	58	59	60
61	62	63	64	65	66	67	68	69	70
71	72	73	74	75	76	77	78	79	80
81	82	83	84	85	86	87	88	89	90
91	92	93	94	95	96	97	98	99	100

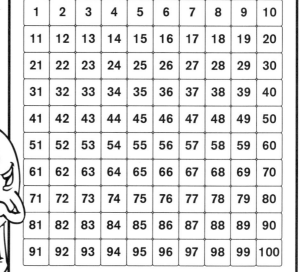

1	2	3	4	5	6	7	8	9	10
11	12	13	14	15	16	17	18	19	20
21	22	23	24	25	26	27	28	29	30
31	32	33	34	35	36	37	38	39	40
41	42	43	44	45	46	47	48	49	50
51	52	53	54	55	56	57	58	59	60
61	62	63	64	65	66	67	68	69	70
71	72	73	74	75	76	77	78	79	80
81	82	83	84	85	86	87	88	89	90
91	92	93	94	95	96	97	98	99	100

What happens? _____

_____

A pattern may be found when the digits of the multiples of nine are added.

For example; 18 → 1 + 8 = 9, 27 → 2 + 7 = 9, 36 → 3 + 6 = 9, 45 → 4 + 5 = 9 ...

3. Consider the digit sums for the numbers produced when 9 is added to 8 (17, 26, 35, 44 etc.). Refer to grid in Question 1.

   What do you notice? _____

   _____

   _____

4. Calculate the digit sums for the multiples you have shaded on the grids in Question 2.

   What do you notice? _____

   _____

Didax Educational Resources ~ www.didaxinc.com

Multiple Multiple Patterns – 1

Each of the cells in the following 1-100 grids has been divided into fourths.

1. (a) Shade all the even numbers (multiples of two) in the top left hand corner of each cell. Now shade all the multiples of three in the top right corner of each cell.

   (b) Which numbers are shaded twice?

   _____

   _____

   Why?_____

   _____

   (c) Why are some numbers not shaded at all?

   _____

   _____

   _____

1	2	3	4	5	6	7	8	9	10
11	12	13	14	15	16	17	18	19	20
21	22	23	24	25	26	27	28	29	30
31	32	33	34	35	36	37	38	39	40
41	42	43	44	45	46	47	48	49	50
51	52	53	54	55	56	57	58	59	60
61	62	63	64	65	66	67	68	69	70
71	72	73	74	75	76	77	78	79	80
81	82	83	84	85	86	87	88	89	90
91	92	93	94	95	96	97	98	99	100

2. Investigate the multiples of four and six.

   (a) Which numbers are shaded twice?

   _____

   _____

   Why?_____

   _____

   (b) Why are some numbers not shaded at all?

   _____

   _____

1	2	3	4	5	6	7	8	9	10
11	12	13	14	15	16	17	18	19	20
21	22	23	24	25	26	27	28	29	30
31	32	33	34	35	36	37	38	39	40
41	42	43	44	45	46	47	48	49	50
51	52	53	54	55	56	57	58	59	60
61	62	63	64	65	66	67	68	69	70
71	72	73	74	75	76	77	78	79	80
81	82	83	84	85	86	87	88	89	90
91	92	93	94	95	96	97	98	99	100

3. Investigate other pairs of multiples; for example, try seven and eight, three and six, or five and eight. Try to predict which numbers will be shaded twice.

   _____

   _____

   _____

   Explain how you made your prediction.

   _____

   _____

   _____

   Check your predictions on the number grid.

1	2	3	4	5	6	7	8	9	10
11	12	13	14	15	16	17	18	19	20
21	22	23	24	25	26	27	28	29	30
31	32	33	34	35	36	37	38	39	40
41	42	43	44	45	46	47	48	49	50
51	52	53	54	55	56	57	58	59	60
61	62	63	64	65	66	67	68	69	70
71	72	73	74	75	76	77	78	79	80
81	82	83	84	85	86	87	88	89	90
91	92	93	94	95	96	97	98	99	100

Didax Educational Resources ~ www.didaxinc.com

Number Grids

# Multiple Multiple Patterns – 2

Each cell in the 1-100 grid has been divided into fourths.

1. Shade all the multiples of three in the top left corner of each cell.

    Shade all the multiples of four in the top right corner of each cell.

    Shade all the multiples of five in the bottom left corner of each cell.

2. List all the numbers that are not shaded at all.

    _____

    _____

1	2	3	4	5	6	7	8	9	10
11	12	13	14	15	16	17	18	19	20
21	22	23	24	25	26	27	28	29	30
31	32	33	34	35	36	37	38	39	40
41	42	43	44	45	46	47	48	49	50
51	52	53	54	55	56	57	58	59	60
61	62	63	64	65	66	67	68	69	70
71	72	73	74	75	76	77	78	79	80
81	82	83	84	85	86	87	88	89	90
91	92	93	94	95	96	97	98	99	100

3. What do you notice about these numbers? _____

    _____

4. (a) Which number(s) is shaded in three corners?_____

    (b) Why?_____

    _____

5. The grid only goes to 100. Predict which number(s) between 100 and 200 will be

    shaded in three corners. _____

    Explain your reasons for making this prediction. _____

    _____

    (You may want to check your prediction on a 101-200 grid.)

6. Now consider the multiples of 2, 4, 6 and 8. Mark them on the grid by shading a different corner for the multiples of each number.

7. Which numbers have exactly two small squares shaded?_____

    _____

8. Which numbers have exactly three small squares shaded?_____

    _____

9. Which numbers have all four squares shaded?

    _____

10. What do you notice about each of these sets of numbers?_____

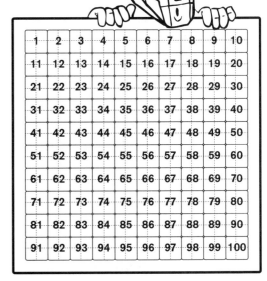

1	2	3	4	5	6	7	8	9	10
11	12	13	14	15	16	17	18	19	20
21	22	23	24	25	26	27	28	29	30
31	32	33	34	35	36	37	38	39	40
41	42	43	44	45	46	47	48	49	50
51	52	53	54	55	56	57	58	59	60
61	62	63	64	65	66	67	68	69	70
71	72	73	74	75	76	77	78	79	80
81	82	83	84	85	86	87	88	89	90
91	92	93	94	95	96	97	98	99	100

Number Grids   Didax Educational Resources ~ www.didaxinc.com

Cracking the Code – 1

There are several different ways to code a message. This coding method requires the use of a blank grid and a numbered grid.

The Method

1. The encoder chooses a grid size and a multiple. For example, using a 100-cell grid and the multiples of 6 allows for a 16 letter message to be sent.

2. Next the encoder needs to choose a number grid, for example, 1-100 and mark in the multiples—in this case, of 6.

3. The encoder then writes his or her 16-letter message onto the blank grid placing each letter in a cell that corresponds to one of the cells previously marked on the 1-100 grid.

4. To finish coding, the encoder fills in the remaining cells with other letters.

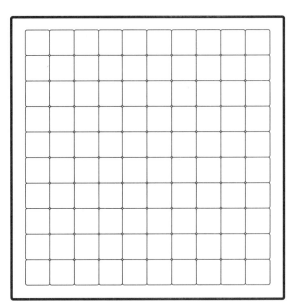

Your Turn

- Make your own coded message using the grids below.

- Pass your message to a friend to decode. You may like to tell him or her which multiple you used and the type of number chart. You can also let your friend try to work out the code without clues.

1	2	3	4	5	6	7	8	9	10
11	12	13	14	15	16	17	18	19	20
21	22	23	24	25	26	27	28	29	30
31	32	33	34	35	36	37	38	39	40
41	42	43	44	45	46	47	48	49	50
51	52	53	54	55	56	57	58	59	60
61	62	63	64	65	66	67	68	69	70
71	72	73	74	75	76	77	78	79	80
81	82	83	84	85	86	87	88	89	90
91	92	93	94	95	96	97	98	99	100

- Experiment with different multiples. How does the length of the message vary? What about using two multiples, for example 3 and 7?

- Start with different number grids, for example, 0-99 or different-sized number grids such as a 1-6 grid.

Cracking the Code – 2

Crack the following codes given the following information.

A	M	C	R	I	L	P	O	B	C
Y	L	E	D	A	S	U	P	E	N
D	R	O	P	N	G	T	I	K	O
H	L	X	C	T	S	O	W	H	W
T	D	S	N	A	B	T	C	O	I
E	U	F	R	T	N	O	A	P	F
A	U	L	E	O	I	G	H	L	R
D	A	N	L	R	D	I	R	A	E
N	O	P	Y	C	R	E	S	I	E
T	V	A	M	S	N	D	Q	O	S

Multiples of 5

H	A	D	R	Q	S	H	O	M	F
S	P	N	E	P	S	A	T	I	M
L	W	M	Y	T	F	L	P	U	L
V	E	A	G	M	C	T	R	U	R
Y	E	R	W	D	M	I	N	W	A
E	B	G	A	K	I	D	U	Q	J
J	U	T	S	M	U	W	O	B	H
V	I	C	R	O	C	T	Y	L	E
H	T	F	H	G	F	J	U	K	D
I	Z	A	L	X	P	M	S	R	N

Multiples of 7

B	E	K	A	R	C	M	E	R	P
A	T	O	P	C	R	T	E	L	R
S	C	A	H	I	D	U	F	M	I
U	R	T	S	N	O	R	U	K	S
R	P	Q	D	E	W	Q	U	G	O
W	J	P	I	O	L	R	O	S	S
N	T	P	K	U	E	T	O	V	J
N	A	Y	N	W	A	Y	M	O	O
P	U	O	N	S	I	H	P	U	L
J	S	Q	U	S	D	Y	S	M	A

Multiples of 4 and 6

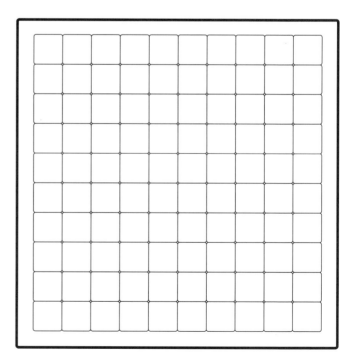

Make a code of your own for a friend to try.

Number Grids

Didax Educational Resources ~ www.didaxinc.com

Digit Sums

When the digits that make up a number are added the result is called the digit sum.

1. Calculate the digit sum for each number in the 1-100 number grid and enter the results into the grid below.

1	2	3	4	5	6	7	8	9	10
11	12	13	14	15	16	17	18	19	20
21	22	23	24	25	26	27	28	29	30
31	32	33	34	35	36	37	38	39	40
41	42	43	44	45	46	47	48	49	50
51	52	53	54	55	56	57	58	59	60
61	62	63	64	65	66	67	68	69	70
71	72	73	74	75	76	77	78	79	80
81	82	83	84	85	86	87	88	89	90
91	92	93	94	95	96	97	98	99	100

For example,
39 → 3 + 9 = 12
12 → 1 + 2 = 3
The digit sum is three.

1	2	3	4	5	6	7	8	9	1
									3
									1

2. Write about any patterns you noticed.

_____

_____

_____

Combining Columns

1. Add a number in the second column to a number in the third column. In which column does the answer appear?

_____

1	2	3	4	5	6	7	8	9	10
11	12	13	14	15	16	17	18	19	20
21	22	23	24	25	26	27	28	29	30
31	32	33	34	35	36	37	38	39	40
41	42	43	44	45	46	47	48	49	50
51	52	53	54	55	56	57	58	59	60
61	62	63	64	65	66	67	68	69	70
71	72	73	74	75	76	77	78	79	80
81	82	83	84	85	86	87	88	89	90
91	92	93	94	95	96	97	98	99	100

2. Investigate adding more numbers in the second column to numbers in the third column. What do you notice?

_____

_____

3. Add numbers in the second and fifth columns. In which column does the answer appear?

_____

_____

4. Now try adding numbers in the second and seventh columns. In which column does the answer appear?

_____

5. Add numbers in the third column to numbers in other columns. Write any patterns that you notice about where the answer appears.

_____

_____

6. Experiment with numbers from different columns and write down what you notice about the column in which the answer appears.

_____

_____

7. Where does the answer appear when a number in a particular column is doubled? Experiment to find out whether any patterns exist.

_____

8. What do you think would happen if a number in a particular column is tripled, quadrupled? In which column would you expect to find the answer?

_____

_____

_____

Didax Educational Resources ~ www.didaxinc.com

Shape Shifter

1. (a) Cut out shape (A) from the bottom of the page. Cover the numbers using the shape in the top left portion of the grid, i.e. 1, 11 and 12.

   (b) Add the numbers.

   _____

1	2	3	4	5	6	7	8	9	10
11	12	13	14	15	16	17	18	19	20
21	22	23	24	25	26	27	28	29	30
31	32	33	34	35	36	37	38	39	40
41	42	43	44	45	46	47	48	49	50
51	52	53	54	55	56	57	58	59	60
61	62	63	64	65	66	67	68	69	70
71	72	73	74	75	76	77	78	79	80
81	82	83	84	85	86	87	88	89	90
91	92	93	94	95	96	97	98	99	100

2. (a) Move the shape one cell to the right to cover 2, 12 and 13.

   (b) Add the numbers.

   _____

3. Continue moving the shape one cell to the right and adding the numbers until you reach 9, 19 and 20. Write the totals for each set of three numbers.

   _____ , _____ , _____ , _____ ,

   _____ , _____ , _____

4. What do you notice about the totals? _____

   _____

5. Place your "L" shape back at the top left of the grid and this time slide the shape one cell down the grid, i.e. 11, 21 and 22 and calculate the total. Continue moving down the grid totalling each set of three numbers until you reach the bottom.

   _____ , _____ , _____ , _____ , _____ , _____ , _____ , _____ , _____ .

6. What do you notice about the totals? _____

   _____

7. What happens if you move in a diagonal path across the grid, i.e. 1, 11, and 12 to 12, 22 and 23 and so on?

   _____

   _____

   _____

   _____

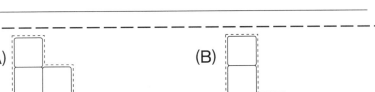

Investigate what happens if you use a different shape, such as the larger "L" shape below or the "T" shape.

 (A)

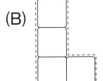 (B)

(C)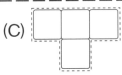

Didax Educational Resources ~ www.didaxinc.com

Number Grids

Windows and Overlays – 1

By cutting holes in number grids, windows can be made to overlay number grids.

1. Make window (A) and then overlay it on the number square. Try the window in different places on the grid. Describe the relationship between the two numbers.

_____

_____

2. What happens if the window is rotated 90°, or a quarter turn? Overlay the window in different places on the grid.

_____

_____

_____

1	2	3	4	5	6	7	8	9	10
11	12	13	14	15	16	17	18	19	20
21	22	23	24	25	26	27	28	29	30
31	32	33	34	35	36	37	38	39	40
41	42	43	44	45	46	47	48	49	50
51	52	53	54	55	56	57	58	59	60
61	62	63	64	65	66	67	68	69	70
71	72	73	74	75	76	77	78	79	80
81	82	83	84	85	86	87	88	89	90
91	92	93	94	95	96	97	98	99	100

3. Make each of the windows and overlaying them on the grid. Try the windows in different places. Explain the effect of each window.

(B) _____

_____

(C) _____

_____

(D) _____

_____

(E) _____

_____

4. What happens when the windows are rotated 90° (or a quarter of a turn) and overlaid on the grid?_____

_____

_____

(A) 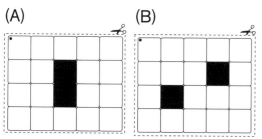 (B) 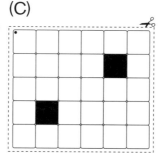 (C) 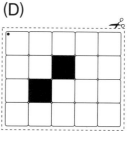 (D) (E)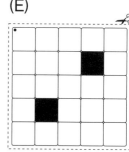

Note: Cut out dark squares.

Windows and Overlays – 2

Windows can be made to overlay on number grids, such as a 1-100 grid. The window shown on the right produces an "add 19" pattern. If you add nineteen to the smaller number, the larger number will be produced. It does not matter where the window is overlaid on the grid.

You could also describe the window as a subtract 19 window!

1	2	3					8	9	10
11	12	13			16		18	19	20
21	22	23					28	29	30
31	32	33		35			38	39	40
41	42	43					48	49	50
51	52	53	54	55	56	57	58	59	60
61	62	63	64	65	66	67	68	69	70
71	72	73	74	75	76	77	78	79	80
81	82	83	84	85	86	87	88	89	90
91	92	93	94	95	96	97	98	99	100

Shade the squares on the grids below to show where you will have to cut to form the correct windows on a 10x10 grid.

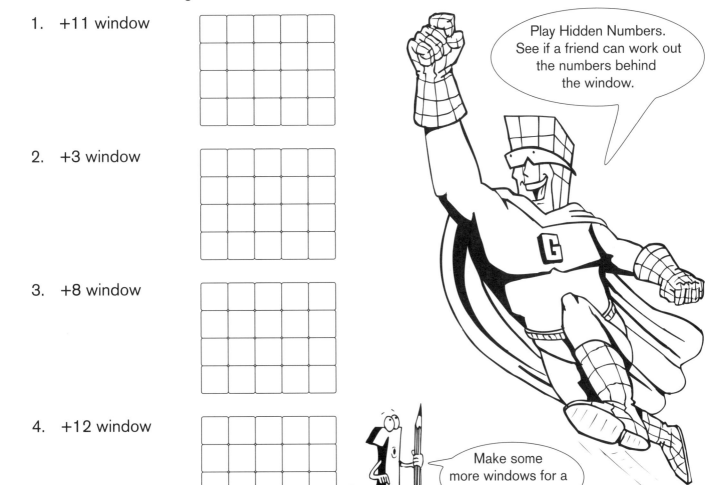

1. +11 window

2. +3 window

3. +8 window

4. +12 window

Play Hidden Numbers. See if a friend can work out the numbers behind the window.

Make some more windows for a friend to try.

Challenge: Make a number window for the multiples of nine. Overlay it on a 1-100 grid, and rotate 90° (or a quarter turn).

What do you notice? _____

Didax Educational Resources ~ www.didaxinc.com

Number Grids

Odd Squares

Draw a 3x3 square anywhere on the 1-100 grid.

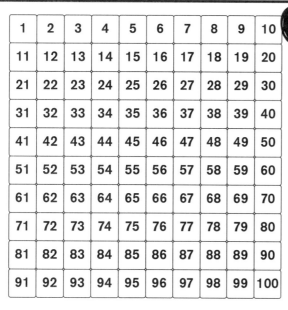

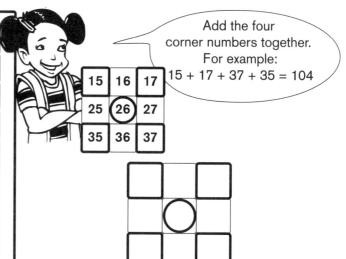

Add the four corner numbers together. For example: 15 + 17 + 37 + 35 = 104

_____ + _____ + _____ + _____ = _____

1. Look for a relationship between the middle number in the square and the total.

   Write about what you notice. _____

   _____

2. (a) Mark several other 3x3 squares on the 1-100 grid and use the boxes below to check whether a similar relationship exists.

   (b) Explain what you found. _____

   _____

3. Try marking 5x5 squares and 7x7 squares on the number grid. Explore the relationship between the sum of the corner numbers and the center number.

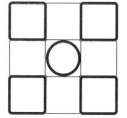   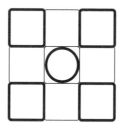   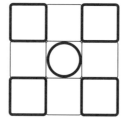   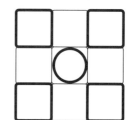

= _____   = _____   = _____   = _____

What happens if you try other grids?

Does the same relationship exist on a multiplication grid?

Number Grids   Didax Educational Resources ~ www.didaxinc.com

Reversals

1. Choose a two-digit number where each of the digits is different (for example, 67).

2. Reverse the number—76.

3. Calculate the difference between the two numbers.

   76 − 67 = _____

4. Shade the square on the number grid.

5. Repeat the process using these numbers. (Put the highest number first.)

   (a) 24   ___42___ − ___24___ = ___18___

   (b) 51   _____ − _____ = _____

   (c) 38   _____ − _____ = _____

   (d) 91   _____ − _____ = _____

   (e) 82   _____ − _____ = _____

   (f) 45   _____ − _____ = _____

1	2	3	4	5	6	7	8	9	10
11	12	13	14	15	16	17	18	19	20
21	22	23	24	25	26	27	28	29	30
31	32	33	34	35	36	37	38	39	40
41	42	43	44	45	46	47	48	49	50
51	52	53	54	55	56	57	58	59	60
61	62	63	64	65	66	67	68	69	70
71	72	73	74	75	76	77	78	79	80
81	82	83	84	85	86	87	88	89	90
91	92	93	94	95	96	97	98	99	100

6. What do you notice about the squares you have shaded?

   _____

   _____

7. Choose six more numbers of your own and note which squares are shaded.

   (a) _____ − _____ = _____

   (b) _____ − _____ = _____

   (c) _____ − _____ = _____

   (d) _____ − _____ = _____

   (e) _____ − _____ = _____

   (f) _____ − _____ = _____

8. Describe what you notice. _____

   _____

9. Explain why you think this is happening. _____

   _____

   _____

Number Grids

Revealing Rectangles

1. Look at the top rectangle on the 1-100 number grid. Lightly shade in the corner numbers.

2. Write each pair of numbers showing the opposite corners and add them.

   14 + 26 = _____, 24 + _____ = _____

   What do you notice about the answers?

   _____

1	2	3	4	5	6	7	8	9	10
11	12	13	14	15	16	17	18	19	20
21	22	23	24	25	26	27	28	29	30
31	32	33	34	35	36	37	38	39	40
41	42	43	44	45	46	47	48	49	50
51	52	53	54	55	56	57	58	59	60
61	62	63	64	65	66	67	68	69	70
71	72	73	74	75	76	77	78	79	80
81	82	83	84	85	86	87	88	89	90
91	92	93	94	95	96	97	98	99	100

3. Two other rectangles are marked on the grid. Add the pairs of numbers in the opposite corners of each of these rectangles.

   (a) 42 + _____ = _____, 44 + _____ = _____

   (b) 78 + _____ = _____, _____ + _____ = _____

   What do you notice about these answers? _____

   _____

4. Draw some rectangles of your own on a 1-100 grid and follow the same steps. What do you notice? _____

   _____

I'm going to try squares.

I'm going to try long skinny rectangles.

I'm going to try small rectangles.

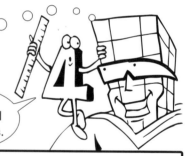

5. Investigate what happens when you draw parallelograms on the 1-100 grid.

   _____

   _____

   _____

6. What happens if you use a 0-99 number grid?

   _____

   _____

   _____

1	2	3	4	5	6	7	8	9	10
11	12	13	14	15	16	17	18	19	20
21	22	23	24	25	26	27	28	29	30
31	32	33	34	35	36	37	38	39	40
41	42	43	44	45	46	47	48	49	50
51	52	53	54	55	56	57	58	59	60
61	62	63	64	65	66	67	68	69	70
71	72	73	74	75	76	77	78	79	80
81	82	83	84	85	86	87	88	89	90
91	92	93	94	95	96	97	98	99	100

Four Squares

1. Several squares have been drawn on the 1-100 grid. Multiply the numbers diagonally opposite each other.

   (a)   12   x   23   = _____ ,

         13   x   22   = _____

   (b)   35   x   46   = _____ ,

         36   x _____ = _____

   (c)   52   x _____ = _____ ,

         _____ x _____ = _____

   (d)   _____ x _____ = _____ ,

         _____ x _____ = _____

   (e)   _____ x _____ = _____ ,

         _____ x _____ = _____

1	2	3	4	5	6	7	8	9	10
11	12	13	14	15	16	17	18	19	20
21	22	23	24	25	26	27	28	29	30
31	32	33	34	35	36	37	38	39	40
41	42	43	44	45	46	47	48	49	50
51	52	53	54	55	56	57	58	59	60
61	62	63	64	65	66	67	68	69	70
71	72	73	74	75	76	77	78	79	80
81	82	83	84	85	86	87	88	89	90
91	92	93	94	95	96	97	98	99	100

2. What do you notice about the answers? _____

1	2	3	4	5	6	7	8	9	10
11	12	13	14	15	16	17	18	19	20
21	22	23	24	25	26	27	28	29	30
31	32	33	34	35	36	37	38	39	40
41	42	43	44	45	46	47	48	49	50
51	52	53	54	55	56	57	58	59	60
61	62	63	64	65	66	67	68	69	70
71	72	73	74	75	76	77	78	79	80
81	82	83	84	85	86	87	88	89	90
91	92	93	94	95	96	97	98	99	100

3. Investigate what happens when you draw squares that are 3 cells by 3 cells.

   (a)   _____ x _____ = _____ ,

         _____ x _____ = _____

   (b)   _____ x _____ = _____ ,

         _____ x _____ = _____

   (c)   _____ x _____ = _____ ,

         _____ x _____ = _____

   (d)   _____ x _____ = _____ ,

         _____ x _____ = _____

4. What do you notice about the answers?

   _____

   _____

5. Try investigating 4x4 squares. What do you notice?

   _____

   _____

   _____

   _____

Notes

Didax Educational Resources ~ www.didaxinc.com

Number Grids

Notes

Notes

About the Author, Paul Swan

Paul Swan has worked as an elementary and high school teacher and now works as a teacher educator at Edith Cowan University, Perth, Western Australia. He was awarded his Ph.D. for his research in children's computation choices and methods.

He has a passion for mathematics and is always looking for ways to make mathematics interesting. He writes mathematics books in his spare time, as well as presenting conference workshops, teacher professional development seminars and mathemagic shows across Australia.

As a father of four boys – all in school – he recognizes the challenges faced by teachers trying to motivate children to learn mathematics. As a result, he tries to make his publications stimulating and inviting for children, while at the same time challenging them to think.